S0-BRN-884

PREFACE

The text that follows on the theory of abelian groups is based on a course given by the author at the University of Chicago during the summer quarter, 1969. For the most part, the class consisted of students beginning their second year of graduate school.

It is not suggested that the ensuing material is a complete treatment of the subject. For this, the reader should consult L. Fuchs' Abelian Groups which will be extended in forthcoming volumes. Other omissions that come to mind are: A. L. S. Corner's dramatic results on endomorphism rings, R. S. Pierce's elaborate discussion of the structure of Hom(A,B), and P. Crawley's and B. Jonsson's results of uniqueness for certain types of decompositions.

The main purpose of this text is to cut through the rudiments of abelian group theory and rather quickly to arrive at P. Hill's version of Ulm's Theorem along with a sampling of R. J. Nunke's homological techniques in dealing with the notion of purity. The discussions of torsion free groups and extensions of groups reflect the author's own taste. The bibliography is not intended to be exhaustive, though it contains many references not mentioned in the text or included in I. Kaplansky's extensive bibliography in his Infinite Abelian Groups.

My thanks are due to those students who not only listened to my lectures, but also added helpful comments. My warmest appreciation goes to Professor Irving Kaplansky for his many timely remarks throughout the course. Miss Sue Podraza has expertly typed the manuscript.

PRELIMINARY FACTS

The text that follows on the theory of abelian (= commutative) groups requires some preliminary notations and concepts. Most of these are somewhat standard (except for new notions in recent papers) and can be found in Kaplansky [69] or Fuchs [32] which are the main two sources of consultation for these notes. This chapter is meant largely for reference and should be referred to as information is needed.

We always use additive notation. In particular, $A + B$ and $\sum A_i$ always mean <u>direct sum</u> while $\{A,B\}$ denotes the <u>group generated by A and B</u>. Square brackets [ ] are used to denote sets. Also ΠA_i denotes the <u>direct product</u> of the groups A_i (called "complete direct sum" by Fuchs [32] and Kaplansky [69] and written $\sum^* A_i$), that is, all vectors $\langle a_i \rangle$ with $a_i \in A_i$ and with componentwise addition. If G and H are groups[1] and if $\phi: G \to X$ and $\psi: H \to X$ are epimorphisms with $A = \text{Ker}\phi$ and $B = \text{Ker}\psi$, then $S = \{(g,h) \in G + H: \phi(g) = \psi(h)\}$ is called the <u>subdirect sum of G and H with kernels A and B, respectively</u>.

For convenience we list some of the groups which are very useful and fundamental in the study of abelian groups. The notation given below for these groups will be standard throughout the text.

(a) $Z \equiv$ additive group of integers.

(b) Cyclic groups $Z(n) = Z/nZ$.

(c) Direct sums of cyclic groups; in particular, free groups (i.e., direct sums of copies of Z).

(d) $\Pi_{\aleph_0} Z \equiv$ Specker group.

---

[1] Hereafter the term "group" always means "abelian group".

2

(e) $Q \equiv$ additive group of rational numbers.

(f) For p a prime, $Z(p^\infty)$ denotes the subgroup of Q/Z generated by cosets of the form $1/p^n + Z$. It is straightforward to show that $Z(p^\infty)$ is an ascending union of subgroups $C_n \simeq Z(p^n)$ and that, consequently, $Z(p^\infty)$ is indecomposable.

(g) One can further show: $Q/Z = \sum_{\text{primes } p} Z(p^\infty)$.

(h) If N is a subset of the primes, I_N denotes the integers <u>localized</u> to the set of primes N, that is, $I_N = \{n/m \; \varepsilon \; Q: (m,p) = 1 \text{ for } p \; \varepsilon \; N\}$.

(i) For $N = [p]$, we use the notation I_p.

(j) $I_p^* \equiv$ p-adic completion of I_p and is often called the <u>group of p-adic integers</u>. (Completions will be taken up in Chapter III.)

(k) If A and B are groups, then Hom(A,B) is the group (abelian) of homomorphisms from A to B with addition being addition of functions.

(l) Let A and B be abelian groups and let F be the free group with pairs (a,b), $a \; \varepsilon \; A$ and $b \; \varepsilon \; B$, as free generators. Let K be the subgroup generated by all elements of the form $(a_1 + a_2,b) - (a_1,b) - (a_2,b)$ and $(a,b_1 + b_2) - (a,b_1) - (a,b_2)$. We define the <u>tensor product of A and B</u> to be $A \otimes B = F/K$ and we denote the elements (a,b) + K by $a \otimes b$. Note that $(a_1 + a_2) \otimes b = a_1 \otimes b + a_2 \otimes b$ and that $a \otimes (b_1 + b_2) = a \otimes b_1 + a \otimes b_2$. Moreover, any $x \; \varepsilon \; A \otimes B$ has a representation (not unique) as a finite sum $x = \sum_i a_i \otimes b_i$. Finally, if f is a bilinear function from $A \otimes B$ into some group H, the mapping $a \otimes b \to f(a,b)$ induces a unique homomorphism from $A \otimes B$ to H.

The reader should have some facility with ordinal and cardinal numbers and with the various forms of Zorn's Lemma (including transfinite induction).

For example, if H is a subgroup of a group G and if K is a subgroup of G with the property that $H \cap K = 0$, then one should be able to deduce from Zorn's Lemma that there is a subgroup M of G maximal with respect to (set inclusion) the properties $K \subseteq M$ and $H \cap M = 0$.

As is the case with many branches of mathematics of this time, we will need some language and notions from the subject of homological algebra. Basic references here are MacLane [88] and Cartan and Eilenberg [14]. With the possible exception of Chapter VI, we make only a mild use of these techniques. For example, we assume an elementary knowledge of exact sequences and commutative diagrams together with the notations $=\!=$, $\rightarrowtail$, $\twoheadrightarrow$ which mean *identity map*, *monomorphism*, and *epimorphism*, respectively. If $A \overset{i}{\rightarrowtail} B \overset{j}{\twoheadrightarrow} C$ is a short exact sequence and if $f: D \to C$ is a homomorphism, we refer to the *pullback of* $A \overset{i}{\rightarrowtail} B \overset{j}{\twoheadrightarrow} C$ *with respect to f* to mean the commutative diagram

$$\begin{array}{ccccc} A & \overset{\tau}{\rightarrowtail} & E & \overset{\sigma}{\twoheadrightarrow} & D \\ \| & & \downarrow{\scriptstyle \phi} & & \downarrow{\scriptstyle f} \\ A & \overset{i}{\rightarrowtail} & B & \overset{j}{\twoheadrightarrow} & C \end{array}$$

where $E = \{(b,d) \in B + D: j(b) = f(d)\}$ and τ, σ and ϕ are defined by $\tau: a \to (a,0)$, $\sigma: (b,d) \to d$ and $\phi: (b,d) \to b$.

If on the other hand we have a homomorphism $f: A \to D$, the terminology *pushout of* $A \overset{i}{\rightarrowtail} B \overset{j}{\twoheadrightarrow} C$ *with respect to f* refers to the commutative diagram

$$\begin{array}{ccccc} A & \overset{i}{\rightarrowtail} & B & \overset{j}{\twoheadrightarrow} & C \\ {\scriptstyle f}\downarrow & & \downarrow{\scriptstyle \psi} & & \| \\ D & \overset{\rho}{\rightarrowtail} & M & \overset{\pi}{\twoheadrightarrow} & C \end{array}$$

where $M = (D + B)/N$, for $N = \{(-f(a), i(a)): a \in A\}$ and where ρ, π and ψ are defined by $\rho: d \to (d,0) + N$, $\pi: (d,b) + N \to j(b)$ and $\psi: b \to (0,b) + N$.

We also assume some familiarity with the derived functors of Hom and

$\otimes$, namely, Ext and Tor, respectively. One should consult [88, pp. 67-76] for a discussion of Ext and [88, pp. 150-153] for a similar discussion of Tor. Specifically, one should know that, for groups A and B, Ext(A,B) is an abelian group whose elements are equivalence classes of short exact sequences $B \rightarrowtail X \twoheadrightarrow A$, where the short exact sequences $B \rightarrowtail X \twoheadrightarrow A$ and $B \rightarrowtail Y \twoheadrightarrow A$ are equivalent if and only if there is a commutative diagram

$$\begin{array}{ccccc} B & \rightarrowtail & X & \twoheadrightarrow & A \\ \| & & \downarrow & & \| \\ B & \rightarrowtail & Y & \twoheadrightarrow & A \end{array}$$

The addition of two equivalence classes (known as the <u>Baer Sum</u> in honor of Reinhold Baer) represented by $B \overset{i}{\rightarrowtail} X \overset{j}{\twoheadrightarrow} A$ and $B \overset{\rho}{\rightarrowtail} Y \overset{\pi}{\twoheadrightarrow} A$, respectively, results in the equivalence class represented by $B \overset{\tau}{\rightarrowtail} W \overset{\sigma}{\twoheadrightarrow} A$, where $W = M/N$, $M = \{(x,y) \in X + Y\colon j(x) = \pi(y)\}$, $N = \{(-i(b),\rho(b))\colon b \in B\}$, $\tau\colon b \to (i(b),0) + N$ and $\sigma\colon (x,y) + N \to j(x)$. Furthermore, if $f\colon A \to B$ is a homomorphism and C is a group, we have induced homomorphisms $f^*\colon \mathrm{Ext}(B,C) \to \mathrm{Ext}(A,C)$ and $f_*\colon \mathrm{Ext}(C,A) \to \mathrm{Ext}(C,B)$. For $\eta \in \mathrm{Ext}(B,C)$ represented by $C \rightarrowtail X \twoheadrightarrow B$, f^* is defined (in terms of representatives) by the following pullback diagram

$$\begin{array}{rccccc} f^*\eta\colon & C & \rightarrowtail & Y & \twoheadrightarrow & A \\ & \| & & \downarrow & & \downarrow f \\ \eta\colon & C & \rightarrowtail & X & \twoheadrightarrow & B \end{array}$$

Dually, f_* is defined by the pushout diagram

$$\begin{array}{rccccc} \eta\colon & A & \rightarrowtail & X & \twoheadrightarrow & C \\ & f \downarrow & & \downarrow & & \| \\ & B & \rightarrowtail & Y & \twoheadrightarrow & C \end{array}$$

Finally, the short exact sequence $A \rightarrowtail B \twoheadrightarrow C$ and the group G induce two exact sequences:

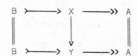

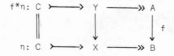

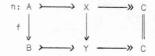

$$(1)\ \mathrm{Hom}(C,G) \rightarrowtail \mathrm{Hom}(B,G) \longrightarrow \mathrm{Hom}(A,G) \xrightarrow{\delta^G}$$

$$\mathrm{Ext}(C,G) \longrightarrow \mathrm{Ext}(B,G) \twoheadrightarrow \mathrm{Ext}(A,G).$$

$$(2)\ \mathrm{Hom}(G,A) \rightarrowtail \mathrm{Hom}(G,B) \longrightarrow \mathrm{Hom}(G,C) \xrightarrow{\delta_G}$$

$$\mathrm{Ext}(G,A) \longrightarrow \mathrm{Ext}(G,B) \twoheadrightarrow \mathrm{Ext}(G,C).$$

For $f \in \mathrm{Hom}(A,G)$, the connecting homomorphism δ^G is defined by the pushout diagram

$$\begin{array}{rcccc} & A & \rightarrowtail & B & \twoheadrightarrow & C \\ f & \downarrow & & \downarrow & & \| \\ \delta^G(f): & G & \rightarrowtail & X & \twoheadrightarrow & C \end{array}$$

and dually δ_G is defined by the corresponding pullback diagram.

Let A and B be groups. Then Tor(A,B) has as generators all symbols $\langle a,m,b\rangle$ with $ma = 0 = mb$, subject to the relations:

(1) $\langle a_1 + a_2,m,b\rangle = \langle a_1,m,b\rangle + \langle a_2,m,b\rangle$, where $ma_1 = 0 = ma_2$

(2) $\langle a,m,b_1 + b_2\rangle = \langle a,m,b_1\rangle + \langle a,m,b_2\rangle$, where $mb_1 = 0 = mb_2$

(3) $\langle a,mn,b\rangle = \langle ma,n,b\rangle$, where $mna = 0 = nb$

(4) $\langle a,mn,b\rangle = \langle a,m,nb\rangle$, where $ma = 0 = mnb$.

An easy consequence of these relations is that $\mathrm{Tor}(A,B) \simeq \mathrm{Tor}(B,A)$. If $f\colon A \to C$ is a homomorphism, there is an induced homomorphism $f : \mathrm{Tor}(A,B) \to \mathrm{Tor}(C,B)$ defined by $f : \langle a,m,b\rangle \to \langle f(a),m,b\rangle$. If $A \rightarrowtail B \xrightarrow{\nu}\!\!\!\twoheadrightarrow C$ is a short exact sequence and G a group, there is an exact sequence

$$\mathrm{Tor}(A,G) \rightarrowtail \mathrm{Tor}(B,G) \longrightarrow \mathrm{Tor}(C,G) \xrightarrow{\partial_G} A \otimes G \longrightarrow B \otimes G \twoheadrightarrow C \otimes G$$

where the connecting homomorphism ∂_G is defined by: For $\langle c,m,g\rangle \in \mathrm{Tor}(C,G)$, $\partial_G\colon \langle c,m,g\rangle \to mb \otimes g$ where $\nu(b) = c$.

Also of importance is the adjoint relation $\mathrm{Hom}(A \otimes B,C) \simeq \mathrm{Hom}(A,\mathrm{Hom}(B,C))$ which induces a natural isomorphism Ext(Tor(A,B),C) Ext(A,Ext(B,C)). The reader is referred to the Appendix for a proof of this relation as well as two theorems of Nunke [102] which involve the functors Hom, Ext, $\otimes$ and Tor and which are needed for Chapter VI.

I. REDUCTION THEOREMS

In this chapter we reduce the study of groups to some extent to the study of groups with special properties.

Definition. For G a group, define $tG = \{x \in G: nx = 0 \text{ for some } n \neq 0\}$. We call tG the torsion subgroup of G (tG is indeed a subgroup of G). A group G is called torsion free if $tG = 0$ and is called torsion if $G = tG$.

Definition. A group M is an extension of a group A by a group B if M contains a subgroup $A' \simeq A$ such that $M/A' \simeq B$, that is, there is a short exact sequence $A \rightarrowtail M \twoheadrightarrow B$.

Theorem 1. Every group G is an extension of a torsion group by a torsion free group.

Proof. Observe that tG is torsion, G/tG is torsion free and that $tG \rightarrowtail G \twoheadrightarrow G/tG$ is exact.

Definition. An element g in a group G is said to be divisible by $n \in Z$ if $g = nx$ has a solution $x \in G$. A group D is called divisible if each element in D is divisible by every positive integer. For a prime p, D will be called p-divisible if each element of D is divisible by p^n for each positive integer n.

Definition. A subgroup H of a group G is an absolute direct summand of G if $G = H + K$ for any subgroup K of G maximal with respect to $H \cap K = 0$.

Proposition 2. A subgroup which is divisible is an absolute direct summand of a group.

Proof. Let G be a group and let D be a divisible subgroup of G (A word of caution is needed here: D a divisible subgroup of G means that for each $d \in D$ and $n \neq 0 \in Z$ there is $d_1 \in D$ such that $d = nd_1$ and not merely $x \in G$ with $d = nx$). Choose H maximal in G with respect to $H \cap D = 0$. If $g \in G - H$, then $\{H,g\} \cap D \neq 0$ by maximality of H, that is,

$h + ng = d \neq 0 \in D$, for some $h \in H$. Note that $n \neq 0$ since $H \cap D = 0$. Therefore, $ng \in \{H,D\}$ from which it follows that $G/\{H,D\}$ is a torsion group.

Suppose that $G/\{H,D\} \neq 0$. It follows that there is $x \in G$ such that $x + \{H,D\}$ has order p for some prime p, i.e., $px = h + d$ for $h \in H$ and $d \in D$. Since D is divisible, $d = pd_1$ where $d_1 \in D$. Then, for $z = x - d_1$, we have that $z + \{H,D\} = x + \{H,D\}$ and so $z + \{H,D\}$ has order p in $G/\{H,D\}$. Since $z \in G - H$, it follows (as above) that $h_1 + nz = d_2 \neq 0 \in D$ where $h_1 \in H$. Moreover, p divides n, say $n = pm$, since $n(z + \{H,D\}) = 0 + \{H,D\}$. Therefore, $0 \neq d_2 = h_1 + nz = h_1 + m(pz) = h_1 + m(px - pd_1) = h_1 + m(h + d - d) = h_1 + mh \in H \cap D$ which contradicts $H \cap D = 0$. Hence, $G = \{H,D\}$ which shows that $G = H + D$.

Definition. A group G is called reduced if its only divisible subgroup is zero.

Theorem 3. Every group G has a unique maximal divisible subgroup D such that $G = G_o + D$ where G_o is reduced.

Proof. Let D be the unique subgroup of G generated by all divisible subgroups of G. If $x \in D$, then $x = x_1 + \dots + x_m$, where each x_i lies in a divisible subgroup of G. So if $n \neq 0 \in Z$, then $x_i = ny_i$ where $y_i \in D$ and, hence, $x = n(\sum_{i=1}^{m} y_i)$ where $y = y_1 + \dots + y_m \in D$. Therefore, D is divisible from which it follows that $G = G_o + D$ by Proposition 2. Clearly, from the definition of D, G_o can have no nonzero divisible subgroups.

Since divisible groups enjoy an elementary structure theory (as will be seen), the study of abelian groups is centered around reduced ones. We have still a further simplification for torsion groups.

Definition. We define $tG_p = \{x \in tG\colon p^n x = 0 \text{ for some } n \geq 0\}$ for each prime p and we say that G is p-primary or a p-group (or simply primary if the prime p is unimportant) if $G = tG_p$. If G is already torsion, then $G_p \equiv tG_p$.

Theorem 4. If G is a torsion group, then G has a unique primary decomposition $G = \sum_{\text{primes } p} G_p$.

Proof. If $x \in G$ and $x \neq 0$, then the order of x has prime decomposition $p_1^{r_1} \ldots p_t^{r_t}$ where the p_i's are pairwise distinct. Let $n = p_1^{r_i} \ldots p_t^{r_i}$ and $n_i = n/p_i^{r_i}$ for $i = 1,\ldots,t$. Clearly, the greatest common divisor of $n_1,\ldots,n_t$ is 1. So there are integers $s_1,\ldots,s_t$ such that $s_1n_1 + \ldots + s_tn_t = 1$. Hence, $x = s_1(n_1x) + \ldots + x_t(n_tx)$ and clearly $n_ix \in G_{p_i}$. Therefore, $G = \{G_p\colon p \text{ a prime}\}$. If $x = x_1 + \ldots + x_t = y_1 + \ldots + y_t$ where $y_i \in G_{p_i}$, then the order of $x_1 - y_1 = (y_2 - x_2) + \ldots + (y_t - x_t)$ divides $p_1^{e_1}$ and $p_2^{e_2} \ldots p_t^{e_t}$ for some $e_1,\ldots,e_t$. It follows that $x_1 - y_1$ has order 1, that is, $x_1 - y_1 = 0$ or $x_1 = y_1$. Continuing in this manner we obtain that $x_i = y_i$ for $i = 1,\ldots,t$. Thus, $G = \sum_{\text{primes } p} G_p$.

Exercises

1. Show that a homomorphic image of a divisible group is divisible. In particular, direct summands of divisible groups are divisible.

2. Prove that a p-group is q-divisible for each prime $q \neq p$.

3. For a group G, define $G' = \bigcap_{n>0} nG$ where $nG = \{g \in G\colon g = nx, x \in G\}$. If G is torsion free, prove that G is divisible and that $G = H + G'$, where $H' = 0$. We shall see that the above result is false when G is a torsion group and is the reason for our note of caution in the proof of Proposition 2.

4. Show that G is p-divisible if and only if $G = pG$ (see Exercise 3 for definition of pG) and that G is divisible if and only if G is p-divisible for each prime p.

5. Prove that if $mx = ng$ has a solution x in the group G where m and n are relatively prime, then $my = g$ has a solution y in G.

6. Let $A \simeq Z(p^\infty)$ and let $a_0 \neq 0 \in A$. Suppose that H is a reduced p-group and that K is a subgroup of H such that $H/K \simeq A/\{a_0\}$. Let G be the subdirect sum of H and A with kernels K and $\{a_0\}$, respectively. Prove that G is reduced.

II. PURITY, BASIC SUBGROUPS, BOUNDED GROUPS' AND FINITELY GENERATED GROUPS

We now undertake a discussion of one of the most fundamental concepts in abelian group theory, namely, that of purity.

Definitions. (1) For $n \in Z$, $nG = \{g \in G: g = nx$, for some $x \in G\}$.

(2) H is a pure subgroup of G if $H \cap nG = nH$ for all $n \in Z$.

(3) H is a p-pure subgroup of G if $H \cap p^nG = p^nH$ for all $n \geq 0$.

If G is a group, then any direct summand, any divisible subgroup and tG are examples of pure subgroups of G (see the exercises at the end of this section). A sufficient (but not necessary) condition for a subgroup H of a group G to be pure is that G/H be torsion free. Nontrivial examples shall soon appear. We further remark that H is a pure subgroup of G if and only if H is a p-pure subgroup for each prime p (see Exercise 4).

Lemma 5. Let A be a group, B a subgroup of A and C a subgroup of B.

(i) If C is pure (p-pure) in B and B is pure (p-pure) in A, then C is pure (p-pure) in A.

(ii) If C is pure (p-pure) in A and B/C is pure (p-pure) in A/C, then B is pure (p-pure) in A.

(iii) If C and B are both pure (p-pure) in A, then B/C is pure (p-pure) in A/C.

(iv) If $B = \bigcup_{i \in I} B_i$ is an ascending union of pure (p-pure) subgroups of A, then B is pure (p-pure) in A.

Proof. We carry out the proofs only for purity and leave the cases of p-purity to the reader.

(i) If $na = c \in C \subseteq B$, then $na = nb$ since B is pure in A. But $nb = c \in C$ implies that $nb = nc_1$, $c_1 \in C$, since C is pure in

B. Thus, $na = nc$ and $C \cap nA = nC$.

(ii) Suppose that $na = b \in B$. Then $na + C = b + C = nb_1 + C$ since B/C is pure in A/C. Therefore, $na = nb_1 + c$ with $b_1 \in B$ and $c \in C$. This implies that $n(a - b_1) = c = nc_1$, $c_1 \in C$, since C is pure in A. Hence, $na = n(b_1 + c_1)$ where $b_1 + c_1 \in B$. Thus, $B \cap nA = nB$.

(iii) If $n(a + C) = b + C$, then $na = b + c = nb_1$ since $b + c \in B$ and B is pure. It follows that $n(a + C) = n(b_1 + C)$ and B/C is pure in A/C.

(iv) If $na = b \in B$, then $b \in B_i$ for some $i \in I$. Therefore, $na = nb_1$, $b_1 \in B_i$, since B_i is pure. Hence, $B \cap nA = nB$ for each $n \in Z$ and so B is a pure subgroup of A.

<u>Definitions</u>. (1) We say that a subset X of a group G is <u>independent</u> if, for distinct elements $x_1, \ldots, x_r \in X$ with $n_1x_1 + \ldots + n_rx_r = 0$, then $n_1x_1 = \ldots = n_rx_r = 0$.

(2) An independent subset X of a group G is called <u>pure independent</u> if $\sum_{x \in X}\{x\}$ is a pure subgroup of G.

(3) An independent subset X of a group G is called <u>p-pure independent</u> if the order of each $x \in X$ is a power of p or infinite and $\sum_{x \in X}\{x\}$ is a p-pure subgroup of G.

<u>Definition</u>. Let p be a prime and G a group. We call a subgroup B_p of G a <u>p-basic subgroup</u> if the following three conditions are satisfied:

(1) B_p is a direct sum of cyclic groups with tB p-primary.

(2) B_p is p-pure in G.

(3) G/B_p is p-divisible.

If G is a p-group, we drop the subscript "p" in B_p and observe from Exercise 3 and Exercise 4 (Chapter I) that (2) and (3) may be replaced by (2') and (3'), respectively, below:

(2') B is pure in G.

(3') G/B is divisible.

A group G may have many p-basic subgroups as will be apparent in later discussion.

Lemma 6. A pure torsion cyclic subgroup of a group G is a direct summand.

Proof. Suppose that $C = \{a\}$ is a pure torsion cyclic subgroup of G of order n. Then clearly $nG \cap C = nC = 0$, since C has order n. Choose H a subgroup of G maximal with respect to $H \supseteq nG$ and $H \cap C = 0$. Suppose that $G \neq \{H,C\}$. As in the proof of Proposition 2, $G/\{H,C\}$ is necessarily a torsion group and contains an element $g + \{H,C\}$ of order p for some prime p. Since $nG \subseteq \{H,C\}$, it follows that p divides n, say $n = pm$. Therefore, $h + pg = ta \in C$ where $h \in H$. Hence, $mh + mpg = mh + ng = mta$ which implies that $mta \in C \cap H = 0$. Therefore, n divides mt and, hence, p must divide t, say $t = ps$. Let $z = g - sa$. Then $z \in G - H$ and so $\{H,z\} \cap C \neq 0$ by choice of H. It follows that $h_1 + rz = c \neq 0 \in C$ and that $r = r_1 p$ since $z + \{H,C\} = g + \{H,C\}$ has order p in $G/\{H,C\}$. But this implies that $0 \neq c = h_1 + rz = h_1 + r_1 pz = h_1 + r_1(pg - psa) = h_1 + r_1(ta - h - ta) = h_1 - r_1 h \in C \cap H$ which contradicts our choice of H.

Lemma 7. (a) If B is a p-pure subgroup of A and A/B is a direct sum of cyclic groups with t(A/B) p-primary, then B is a direct summand of A.

(b) If B is a pure subgroup of A and A/B is a direct sum of cyclic groups, then B is a direct summand of A.

Proof. We shall only prove (b) since (a) has a similar proof. Suppose that $A/B = \sum_{i \in I}\{y_i + B\}$. If $y_i + B$ has finite order n_i in A/B, then $n_i y_i = n_i b_i$, $b_i \in B$, since B is pure in A. Therefore, $(y_i - b_i) + B = y_i + B$ and $y_i - b_i$ has order n_i in A. It follows that we may choose representatives $[x_i]_{i \in I}$ in A such that $x_i + B = y_i + B$ and that x_i has the same order in A as does $x_i + B$ in A/B. Let H be the subgroup of A generated by the set $[x_i]_{i \in I}$. Clearly, $A = \{B,H\}$. Suppose that $b \in H \cap B$. By relabeling subscripts if necessary, we may suppose that $b = m_1 x_1 + \ldots + m_r x_r$. This implies that $b + B = 0 + B = (m_1 x_1 + B) + \ldots + (m_r x_r + B)$ and, therefore, that n_i divides m_i, n_i being the order of x_i (otherwise, $m_i = 0$ if x_i has infinite order). By our choice of $[x_i]_{i\varepsilon I}$, it follows that $m_i x_i = 0$ for $i = 1,\ldots,r$ and, hence, that $b = 0$. Thus, $A = B + H$.

Definition. If G is a group, we define $G[n] = \{x \in G\colon nx = 0\}$. If p is a prime and G is a p-group, we call G[p] the socle of G.

The lemma that follows provides our first glimpse of how the behavior of the socle of a p-group affects the behavior of the entire group.

Lemma 8. If G is a p-group such that each element of its socle is divisible by p^n for all $n \geq 0$, then G is divisible.

Proof. Since G is a p-group, it is enough to prove that G is p-divisible by Exercises 2 and 4 (Chapter I). Furthermore, by Exercise 4 (Chapter I), it is enough to show $G = pG$. Since by hypothesis $G[p] \subseteq pG$, we proceed by induction and suppose that $G[p^n] \subseteq pG$ for $m \leq n$. Let $g \in G[p^{n+1}]$. By hypothesis $p^n g = p^{n+1}x$ for some $x \in G$ since $p^n g \in G[p]$. Therefore, $g = px + a$ where $p^n a = 0$, that is, $a \in G[p^n]$. By our induction hypothesis, $a = pb$ for $b \in G$. Hence, $g = p(x + b) \in pG$. Since $G = \bigcup_{n<\omega} G[p^n]$, it follows that $G = pG$ and G is divisible.

Lemma 9. Every group G such that $pG \neq G$ (that is, G is not p-divisible) has a nonzero p-pure cyclic subgroup C such that C is p-primary or has infinite order.

Proof. First suppose that $p(tG_p) \neq tG_p$. By Lemma 8, we may choose an element a of order p in tG_p such that there is a largest integer $n \geq 0$ for which there is a solution to the equation $p^n x = a$ for $x \in tG_p$, say $p^n b = a$ and let $B = \{b\}$. Suppose that B is not p-pure in tG_p. Then $p^e g = mb$ where $g \in tG_p$ and $m = p^r q$ with p and q relatively prime and $e > r$, $n+1 > r$. Let $k = n + 1 - r$. Then $p^{e+k-1} g = qp^n b = qa$, where $e + k - 1 = e + n + 1 - r - 1 = e + n - r > n$. But by Exercise 5 (Chapter I) $p^{e+k-1} x = a$ also has a solution $x \in G$ which contradicts our choice of a. Hence, B must be p-pure in tG_p and thus B is p-pure in G since tG_p is easily seen to be p-pure (in fact pure) in G.

Secondly, suppose that $p(tG_p) = tG_p$. By Exercises 2 and 4 (Chapter I), tG_p is divisible and so $G = H + tG_p$ where $tH_p = 0$ and $pH \neq H$. Choose $b \in H - pH$ and let $B = \{b\}$. Suppose that $p^n h = mb \in B$, where $m = p^r q$ with p and q relatively prime. If $n > r$, then $p^{n-r}h = qb$, since $tH_p = 0$, which implies that $px = b$ has a solution $x \in H$ by Exercise 5 (Chapter I). However, this latter fact would contradict our choice of b, so $r \geq n$. Hence, B is p-pure in H and thus must also be p-pure in G.

Lemma 10. Let $B = \sum_{i \in I}\{b_i\}$ be a subgroup of G such that tB is p-primary. Then the following two statements are equivalent.

(a) $[b_i]_{i \in I}$ is a maximal p-pure independent subset of G.

(b) B is a p-basic subgroup of G.

Proof. (a) $\Longrightarrow$ (b). We need only show that G/B is p-divisible, that is, $G/B = p(G/B)$. If $G/B \neq p(G/B)$, there is, by Lemma 9, a nonzero cyclic subgroup C/B of G/B which is p-pure and t(C/B) is p-primary. By Lemma 7(a), $C = B + \{x\}$ and, by Lemma 5(ii), C is again p-pure in G. But this latter statement implies that $[b_i]_{i \in I} \cup [x]$ is p-pure independent; contradicting the maximality of $[b_i]_{i \in I}$. Thus, G/B is necessarily p-divisible.

(b) $\Longrightarrow$ (a). If for some $x \in G$ with either infinite order or order a power of p, the set $[b_i]_{i \in I} \cup [x]$ is p-pure independent, then $\{x\} \simeq \{x,B\}/B$ is p-pure in the p-divisible group G/B. But this is impossible since x cannot be divisible by p. Thus, $[b_i]_{i \in I}$ is a maximal p-pure independent subset of G.

We are now in a position to prove the fundamental theorem concerning p-basic subgroups (basic subgroups if our group is p-primary) of a group. This result is due to Kulikov [80] and Fuchs [36].

Theorem 11. Every group G has a p-basic subgroup and any two p-basic subgroups of G are isomorphic.

Proof. If G is p-divisible, then the zero subgroup of G is a p-basic subgroup. Otherwise we may apply Zorn's Lemma to obtain a maximal p-pure independent subset $[b_i]_{i \in I}$. By Lemma 10, $B = \sum_{i \in I}\{b_i\}$ is a p-basic subgroup of G. Write $B = B_0 + B_1 + \dots + B_n + \dots$ where B_0 is free and $B_n \simeq \sum Z(p^n)$ for $n \geq 1$. It is easy to see that $tB = B_1 + \dots + B_n + \dots$ is a p-basic subgroup of tG_p. Since tG_p/tB is p-divisible, we have that $tG_p \simeq \{tB, p^k tG_p\}$ for each integer $k \geq 1$. Hence, $tG_p/p^k tG_p \simeq \{tB, p^k tG_p\}/p^k tG_p \simeq tB/tB \cap p^k tG_p \simeq tB/p^k tB$, since tB is pure in tG_p. The number of copies of $Z(p^n)$ in B_p, $n \geq 1$, is equal to the number of copies of $Z(p^n)$ in $tB/p^{n+1}tB \cong tG_p/p^{n+1}tG_p$. Hence, the torsion subgroups of any two p-basic subgroups are isomorphic.

It is routine to show that $B_0 \simeq \{B_0, tG_p\}/tG_p$ is a p-basic subgroup of G/tG_p (Exercise 8) and in a similar fashion as above that $B_0/pB_0 \simeq (G/tG_p)/p(G/tG_p)$. This shows that the number of copies of Z in B_0 is the same for any two p-basic subgroups, which completes our proof.

Definition. A group B is called bounded if $nB = 0$ for some nonzero integer n.

Theorem 12. A bounded group is a direct sum of cyclic torsion groups.

Proof. A bounded group is necessarily torsion and, since torsion groups have a primary decomposition (Theorem 4, Chapter I), it is enough to prove the above theorem for a bounded p-primary group G, say $p^n G = 0$. Let B be a basic subgroup of G. Then $G/B = p^n(G/B) = 0$ since G/B is divisible and $p^n G = 0$. Thus, $G = B$.

Corollary 13. If G is a torsion group such that G_p is bounded for each prime p, then G is a direct sum of cyclic groups.

Bounded groups possess another attractive feature which was hinted in Lemma 6.

Theorem 14. A bounded pure subgroup of a group is a direct summand.

The proof of this result is left as an exercise (see Exercise 6) since its proof so closely follows that of Lemma 6. Before concluding this chapter with the classical characterization of finitely generated groups, we establish a well-known result concerning free groups.

Theorem 15. A subgroup of a free group is free.

Proof. Let $F = \sum_{\alpha<\gamma}\{x_\alpha\}$ be a free group and suppose that A is a subgroup of F. Define $A_\beta = A \cap \sum_{\alpha<\beta}\{x_\alpha\}$ for each $\beta < \gamma$. If β is a limit ordinal, then it is easily seen that $A_\beta = \bigcup_{\lambda<\beta} A_\lambda$; in particular, $A = \bigcup_{\lambda<\gamma} A_\lambda$. Since $A_{\beta+1}/A_\beta$ is isomorphic to a subgroup of $\{x_\beta\}$, that is, $A_{\beta+1}/A_\beta \simeq 0$ or Z, then by Lemma 7(b), $A_{\beta+1} = A_\beta + C_\beta$, where $C_\beta \simeq 0$ or Z. With $C_0 = A_1$, the above facts show that $A = \sum_{0\le\beta<\gamma} C_\beta$. Thus, A is free.

Theorem 16. A finitely generated group G is a direct sum of cyclic groups.

Proof. First suppose that A is a finitely generated torsion free group. If A is generated by one element, then clearly $A \simeq Z$. To show that, in general, A is free we proceed by induction on the number of generators needed to generate A. So suppose the result holds if a torsion free group needs $\le n$ generators and suppose that A is a torsion free group generated by n+1 elements. Let B be a subgroup of A generated by n elements of a generating set for A consisting of n+1 elements. By the induction hypothesis B is free. Furthermore, A/B is cyclic. If $A/B \simeq 0$ or Z, then $A \simeq B + C$, where $C \simeq 0$ or Z, by Lemma 7(b). Therefore, A is free in this case. Otherwise, $A/B \simeq Z(r)$, for $r > 0$, which implies that $A \simeq rA \subseteq B$ and, hence, that A is free by Theorem 15.

Since G/tG is clearly finitely generated, then from above G/tG is free which implies, by Lemma 7(b), that $G = tG + F$ where F is free. Moreover, tG is finitely generated and, therefore, tG is necessarily bounded. By Theorem 12, tG is a direct sum of cyclic groups.

Exercises

1. Prove that every group is an extension of a direct sum of cyclic groups by a torsion group.

2. Prove that H is a pure subgroup of G if either (i) H is a direct summand of G or (ii) G/H is torsion free.

3. If G is q-divisible, i.e., $qG = G$, for every prime $q \neq p$, then H is a pure subgroup of G if and only if H is a p-pure subgroup of G.

4. Prove that H is a pure subgroup of G if and only if H is a p-pure subgroup of G for every prime p.

5. (Baer) Let G be a p-group and let B be a basic subgroup of G. Write $B = \sum_{n<\omega} B_n$ where $B_n \simeq \sum Z(p^n)$ for each n. Prove that $G = B_1 + \ldots + B_n + \{\sum_{n<m} B_m, p^n G\}$ for each n.

6. Prove Theorem 14.

7. Let G be a p-group such that $\bigcap_{n<\omega} p^n G = \bigcap_{n<\omega} nG = 0$. Show that if F is a finite subgroup of G, then G = A + H where $F \subseteq A$ and A is finite (look at proof of Lemma 9).

8. Let G be a group and let B be a p-basic subgroup of G. Write $B = B_o + tB$ and prove that $B_o \simeq \{B_o, t_p G\}/t_p G$ is a p-basic subgroup of $G/t_p G$.

9. This exercise offers a slicker method for proving Lemma 9. However, one should note that the proof of Lemma 9 effectively proves Exercise 7.

Let G be a group such that $G \neq pG$. Choose $z \in G - pG$ having smallest order (infinite if necessary). Prove that {z} is a p-pure subgroup of G.

This imaginative idea was communicated to me by Peter Kohn; one of the students who patiently made sense of my lectures.

III. PROJECTIVES, INJECTIVES, PURE PROJECTIVES AND PURE INJECTIVES

The ideas of this chapter go back to Baer, 1934, and have recently been formalized with the advent of homological algebra.

Definition. (1) A group P is called projective if, for any epimorphism $\nu: B \twoheadrightarrow C$ and any homomorphism $f: P \to C$, there is a homomorphism $\phi: P \to B$ such that $\nu\phi = f$, that is, there is a commutative diagram

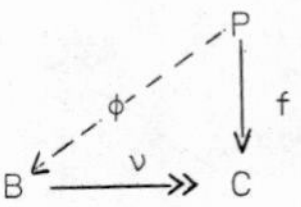

(2) We call a group D injective if dually, for any monomorphism $\pi: A \rightarrowtail B$ and any homomorphism $f: A \to D$, there is a homomorphism $\phi: B \to D$ such that $f = \phi\pi$, that is, there is a commutative diagram

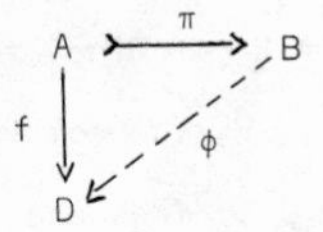

Our first theorem of this chapter says (in view of Theorems 18 and 19 yet to come), in the language of homological algebra, that there are "enough" projectives and injectives.

Theorem 17. Every group is a homomorphic image of a free group and, in addition, every group can be embedded in a divisible group.

Proof. Let G be a group and form the free group $F = \sum_{g\epsilon G}\{x_g\}$, indexed by G, and define $\phi: F \to G$ be defining $\phi(x_g) = g$ on the basis $[x_g]_{g\epsilon G}$ of F. Clearly, ϕ is an epimorphism. Moreover, F can be embedded as a subgroup of a direct sum D of copies of Q (one copy of Q for each $g \in G$). Then $G \simeq F/\mathrm{Ker}\phi \subseteq D/\mathrm{Ker}\phi = D_1$ and D_1 is divisible since it is a homomorphic image of D.

Our next two theorems characterize projectivity and injectivity in abelian groups.

Theorem 18. Let G be a group. Then the following are equivalent:

(1) G is projective.

(2) Every short exact sequence $A \rightarrowtail B \twoheadrightarrow G$ splits.

(3) G is free.

Proof. (1) $\Longrightarrow$ (2). If $A \overset{i}{\rightarrowtail} B \overset{\nu}{\twoheadrightarrow} G$ is short exact and if G is projective, then there is a commutative diagram

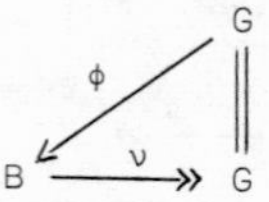

Clearly, $\mathrm{Image}\phi \cap \mathrm{Ker}\nu = 0$ and $B = \{\mathrm{Ker}\nu, \mathrm{Image}\phi\}$. Hence, $\mathrm{Ker}\nu = i(A)$ is a direct summand of B and so $A \overset{i}{\rightarrowtail} B \overset{\nu}{\twoheadrightarrow} G$ splits.

(2) $\Longrightarrow$ (3). By Theorem 17, there is an exact sequence $K \rightarrowtail F \twoheadrightarrow G$ with F free. By hypothesis $K \rightarrowtail F \twoheadrightarrow G$ splits and so G is isomorphic to a direct summand of a free group. By Theorem 15, G is free.

(3) $\Longrightarrow$ (2). Let $G = \sum_{i \in I} \{x_i\}$ be free and $\nu: B \twoheadrightarrow C$ and $f: G \to C$ homomorphisms. Define $\phi: G \to B$ by defining $\phi(x_i) = b_i$ where b_i is chosen such that $\nu(b_i) = f(x_i)$. Clearly, $\nu\phi = f$.

Theorem 19. Let G be a group. Then the following are equivalent:

(1) G is injective.

(2) Every short exact sequence $G \rightarrowtail B \twoheadrightarrow C$ splits.

(3) G is divisible.

Proof. (1) $\Longrightarrow$ (2). Let G be injective and $G \overset{\pi}{\rightarrowtail} B \twoheadrightarrow C$ short exact. Since G is injective, there is a commutative diagram

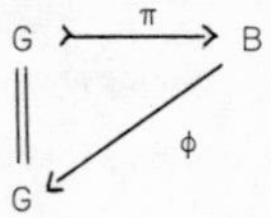

Clearly, $B \simeq \text{Image}\pi + \text{Ker}\phi$ and so $G \overset{\pi}{\rightarrowtail} B \twoheadrightarrow C$ splits.

(2) $\Longrightarrow$ (3). By Theorem 17, there is an exact sequence $G \rightarrowtail D \twoheadrightarrow E$ with D divisible. By hypothesis this exact sequence splits. So G is a direct summand of a divisible group and, therefore, is divisible.

(3) $\Longrightarrow$ (1). We now suppose that G is divisible and $\pi: A \rightarrowtail B$ and $f: A \to G$ homomorphisms. We may suppose that π is the inclusion homomorphism. Consider the collection of all pairs (H,ϕ) such that $A \subseteq H \subseteq B$ and ϕ extends f. We order this collection by: $(H_1,\phi_1) \le (H_2,\phi_2)$ if and only if $H_1 \subseteq H_2$ and ϕ_2 extends ϕ_1. If $[(H_i,\phi_i)]_{i\varepsilon I}$ is a well ordered chain, then (H,ϕ) is an upper bound where $H = \bigcup_{i\varepsilon I} H_i$ and $\phi = \sup_{i\varepsilon I} \phi_i$, i.e., $\phi(x) = \phi_i(x)$ if $x \varepsilon H_i$. Therefore, we apply Zorn's Lemma to obtain a maximal element (H,ϕ). Suppose that $H \neq B$. The maximality of (H,ϕ) implies that G/H is a torsion group (in fact, $\{g\} \cap H \neq 0$ for $g \neq 0 \varepsilon G$). So we may select $b \varepsilon G - H$ such that $pb \varepsilon H$ for some prime p. Since G is divisible, there is an element $z \varepsilon G$ such that $pz = \phi(pb)$. Define $\psi: \{H,B\} \to G$ by $\psi(h + nb) = \psi(h) + nz$. It is straightforward to check that ψ is well defined and is a homomorphism of $\{H,b\}$ into G extending ϕ. This contradicts the maximality of (H,ϕ). Thus, $H = G$ and ϕ extends f.

<u>Definition</u>. A subgroup A of a group B is called an <u>essential subgroup</u> if each nonzero subgroup of B has nonzero intersection with A.

The following result establishes the existence of injective envelopes for abelian groups (see the definition below).

<u>Theorem 20</u>. Every group G can be embedded as an essential subgroup in a divisible group D. The above embedding is unique in the sense that, if G is an essential subgroup of some other divisible group D', there is an isomorphism between D and D' which extends the identity map on G.

Proof. Clearly, there are embeddings $Z \rightarrowtail Q$ and $Z(p^n) \rightarrowtail Z(p^\infty)$ and these embeddings are essential. Hence, any indecomposable cyclic group can be embedded as an essential subgroup of a divisible group. By Zorn's Lemma we may choose a maximal independent subset $[a_i]_{i\varepsilon I}$ in G such that each $\{a_i\}$ is indecomposable. Applying our above discussion, we let D_i be a divisible group containing $\{a_i\}$ as an essential subgroup. Let $D = \sum_{i\varepsilon I} D_i$ and let f denote the monomorphism which embeds $A = \sum\{a_i\}$ coordinatewise into $D = \sum D_i$. Let $d \neq 0 \;\varepsilon\; D$. If d has only one nonzero coordinate, there clearly exists $n \;\varepsilon\; Z$ such that $nd \neq 0 \;\varepsilon\; A$. Suppose the same is true for any element of D having less than or equal to r nonzero coordinates and let $d = x + y$ where x has r nonzero coordinates and y has one nonzero coordinate distinct from those for x. By the induction hypothesis there is an $n_1 \;\varepsilon\; Z$ such that $n_1 x \;\varepsilon\; A - 0$. If $n_1 y \;\varepsilon\; A$, then $n_1 d \;\varepsilon\; A - 0$ and we are finished. If $n_1 y \not\varepsilon\; A$, there is an $n_2 \;\varepsilon\; Z$ such that $n_2 n_1 y \;\varepsilon\; A - 0$ and, therefore, $n_2 n_1 d \;\varepsilon\; A - 0$. In any case, A is an essential subgroup of D.

Since D is injective, there is a homomorphism $\phi: G \to D$ which extends f. Since $\mathrm{Ker}\phi \cap A = 0$ and A is generated by a maximal independent subset of G, it follows that ϕ is monic. Since $A \subseteq \mathrm{Image}\phi \subseteq D$ and since A is essential in D, it follows that $\mathrm{Image}\phi \simeq G$ is essential in D.

Now suppose that G is an essential subgroup of the divisible groups D and D'. By Theorem 19, there is a homomorphism $\psi: D \to D'$ which extends the identity on G. Since $\mathrm{Ker}\psi \cap G = 0$ and G is essential in D, then $\mathrm{Ker}\psi = 0$. Moreover, $\mathrm{Image}\psi$ is divisible and so $D' = \mathrm{Image}\psi + K$ with $G \subseteq \mathrm{Image}\psi$. Hence, $G \cap K = 0$ which implies that $K = 0$. Thus, ψ is an isomorphism.

The group D described in the above theorem is called the <u>minimal divisible group containing G</u> or <u>injective envelope of G</u> which we denote by D(G).

It is clear that Q is indecomposable. Moreover, $Z(p^\infty)$ is indecomposable since its socle is Z(p). Since an indecomposable cyclic group is of the form Z or $Z(p^n)$ and $D(Z) = Q$ and $D(Z(p^n)) = Z(p^\infty)$, examination of the proof of the above theorem shows that we have further proved the following result.

Theorem 21. Every divisible group D is the direct sum of indecomposable divisible groups which are either isomorphic to Q or $Z(p^\infty)$ for various primes p.

One sees, in addition, that the cardinal numbers in the direct sum $D = \sum_{\alpha_0} Q + \sum_p \sum_{\alpha_p} Z(p^\infty)$ form a complete set of cardinal invariants for the class of divisible groups. In Chapter IV we investigate further the topic of cardinal invariants for classes of abelian groups.

Definition. Let G be a group and let $D(G) = \sum_{\alpha_0} Q + \sum_p \sum_{\alpha_p} Z(p^\infty)$ be the injective envelope of G. We define the rank of G by rank $(G) = \alpha_0 + \sum_p \alpha_p$.

We now focus our attention on the relative concepts of pure projectivity and pure injectivity. We begin by introducing a new notion, that of height, and taking a more penetrating look at direct sums of cyclic groups.

Definition. Let G be a p-group and let $x \in G$. We define the height of x in G by:

$$h^G(x) = \begin{cases} n \text{ if } x \in p^n G \text{ but } x \notin p^{n+1} G \\ \infty \text{ if } x \in \bigcap_{n<\omega} p^n G \end{cases}$$

We shall refine this notion in Chapter IV. Also, in later chapters, we may use the notation $h_p^G(x)$ if we must keep track of the prime p. Observe that a subgroup H of a p-group G is pure if and only if $h^G(x) = h^H(x)$ for each $x \in H$. Our first application of the concept of height is the Kulikov Criterion for direct sums of cyclic p-groups. (This result has been generalized by Megibben [99].)

Definition. Define $G' = \bigcap_{n<\omega} nG$. If G is a p-group, then in fact $G' = \bigcap_{n<\omega} p^n G$. We say that a group G has no elements of infinite height provided $G' = 0$ (we ignore zero in G).

In the following theorem, all heights are computed with respect to the containing group unless otherwise stated.

Theorem 22. A p-group G is a direct sum of cyclic groups if and only if the socle of G is an ascending union $G[p] = \bigcup_{n<\omega} S_n$ of subgroups such that the heights of the nonzero elements in S_n remain under a finite bound k_n.

Proof. Since the necessity is clear, we suppose that $G[p] = \bigcup_{n<\omega} S_n$ where $S_n \subseteq S_{n+1}$ and $h^G(x) < k_n$ for each $x \neq 0 \in S_n$. We construct inductively pure independent sets X_n such that $\sum_{x \in X_n} \{x\} \cap G[p] = S_n$ and $X_n \subseteq X_{n+1}$. We may suppose that $S_1 = 0$ and $X_1 = \phi$ and that the construction of X_n is completed satisfying both of the above conditions. Among the pure independent sets Y in G such that $X_n \subseteq Y$ and $\sum_{y \in Y} \{y\} \cap G[p] \subseteq S_{n+1}$, we choose X to be maximal with respect to inclusion. Suppose that $\sum_{x \in X} \{x\} \cap G[p] \subsetneq S_{n+1}$ and let $z \in S_{n+1} - \sum_{x \in X} \{x\}$. Since all elements of the form $z + a$, $a \in \sum_{x \in X} \{x\} \cap G[p]$, have height $< k_{n+1}$, we may choose $w = z + a$ to have maximal height among this collection, say $h^G(w) = r < k_{n+1}$ where $w = p^r b$. Suppose that $H = \{b\} + \sum_{x \in X} \{x\}$ and that $p^m g \in H[p]$. If $p^m g \in \sum_{x \in X} \{x\}$, then $p^m g = p^m h$ for $h \in \sum_{x \in X} \{x\}$, since X is pure independent. If $p^m g = qw + a$, $a \in \sum_{x \in X} \{x\} \cap G[p]$ and q and p are relatively prime, then $qw + a$ and, hence, $w + ua$ have height $\geq m$ in G, where $qu \equiv 1 \pmod p$. By definition of w, it follows that $h^G(w) = r \geq m$ and, therefore, $h^G(ua) \geq m$, since $p^m ug - w = ua$. Hence, $h^G(x) = h^H(x)$ for each $x \in H[p]$. To finish the argument that H is pure, we proceed by induction and suppose that $h^H(x) = h^G(x)$ for each $x \in H[p^m]$. If $h \in H[p^{m+1}]$ and if $p^n g = h$, then $p^{n+1} g = ph \in H[p^m]$ and, therefore, $p^{n+1} g = p^{n+1} h_1$ by the induction hypothesis. Hence, $p^n g = p^n h_1 + y$, $y \in H[p]$. From above $h^G(y) = h^H(y)$ which yields that $y = p^n h_2$. Hence, $p^n g = p^n(h_1 + h_2) = h$ proving that $h^G(h) = h^H(h)$. But $H = \{b\} + \sum_{x \in X} \{x\}$ a pure subgroup of G together with $H \cap G[p] \subseteq S_{n+1}$ contradicts the maximality of X. Thus, $\sum_{x \in X} \{x\} \cap G[p] = S_{n+1}$ and so we set $X_{n+1} = X$.

Therefore, we obtain a pure independent set $P = \bigcup_{n<\omega} X_n$. Let $C = \sum_{x \in P} \{x\}$. Then C is a pure subgroup of G such that $C[p] = G[p]$. Suppose that $G[p^n] \subseteq C$ and let $g \in G[p^{n+1}]$. Then $pg \in C \cap pG = pC$, that is $pg = pc$ or

$g = c + s$, where $s \in G[p] = C[p]$. Hence, $G[p^{n+1}] \subseteq C$ and thus $G = C$ by induction.

We now have our first glimpse of the strength of the hypothesis of countability in the theory of abelian groups.

<u>Corollary 23</u>. (Prüfer) A countable p-group with no elements of infinite height is a direct sum of cyclic groups.

<u>Proof</u>. Write $G[p] = \{x_1, x_2, \ldots, x_n, \ldots\}$ and let $S_n = \{x_1, \ldots, x_n\}$. Then $S_n \subseteq S_{n+1}$, $G[p] = \bigcup_{n<\omega} S_n$ and $\sup[h^G(x)\colon x \in S_n - 0] < \infty$ since S_n is finite and $\bigcap_{n<\omega} p^n G = 0$. Thus, Theorem 22 shows immediately that G is a direct sum of cyclic groups.

<u>Theorem 24</u>. A subgroup of a direct sum of cyclic groups is again a direct sum of cyclic groups.

<u>Proof</u>. Let G be a direct sum of cyclic groups and let H be a subgroup of G. Since G/tG is free, then H/tH is also free by Theorem 15. Therefore, $H = F + tH$ where F is free. Since $tH = \sum_{\text{primes}} tH_p$ and $tH_p \subseteq tG_p$, it is enough to prove the result when G and H are both p-groups for a fixed prime p. Now $G[p] = \bigcup_{n<\omega} S_n$ where $S_n \subseteq S_{n+1}$ and $h^G(x) \leq n$ for $x \in S_n - 0$. Let $T_n = H \cap S_n$. Then clearly $H[p] = \bigcup_{n<\omega} T_n$, $T_n \subseteq T_{n+1}$ and $h^H(x) \leq h^G(x) \leq n$ for $x \in T_n - 0$. Applying Kulikov's Criterion we see that H is a direct sum of cyclic groups.

<u>Definition</u>. (1) A group P is called <u>pure projective</u> if, for any epimorphism $\nu\colon B \twoheadrightarrow C$ with $\operatorname{Ker}\nu$ pure in B and any homomorphism $f\colon P \to C$, there is a homomorphism $\phi\colon P \to B$ such that $\nu\phi = f$.

(2) A group H is called <u>pure injective</u> if, for any monomorphism $\pi\colon A \to B$ with $\operatorname{Image}\pi$ pure in B and any homomorphism $f\colon A \to H$, there is a homomorphism $\phi\colon B \to H$ such that $f = \phi\pi$.

Corresponding to our theorem showing the existence of "enough" projectives and injectives, we have ...

<u>Theorem 25</u>. Every group G is the homomorphic image of a direct sum of cyclic groups with pure kernel and, moreover, G can be embedded as a

pure subgroup of a direct sum $A + D$, where A is a direct product of cyclic torsion groups and D is divisible.

<u>Proof</u>. Let $G = [g_i]_{i\varepsilon I}$, form the direct sum $P = \sum_{i\varepsilon I}\{g_i\}$ and define a homomorphism σ coordinatewise by sending g_i to itself in G. Clearly, σ is an epimorphism. Suppose that $x \varepsilon P$ and $nx \varepsilon \mathrm{Ker}\sigma$, say $x = n_i g_i + \ldots + n_r g_r$ (relabel coordinates if necessary!) as a formal sum in P. Then, for some $i \varepsilon I$, $g_i = \phi(x) = \sum_{j=1}^{r} n_j g_j$ (sum in G) and $ng_i = 0$. Let y be the element in P, $y = x - g_i$. Then plainly $y \varepsilon \mathrm{Ker}\sigma$ and $ny = n(x - g_i) = nx$, since $ng_i = 0$. Thus, $\mathrm{Ker}\sigma$ is a pure subgroup of P.

Let $D = D(G')$ and let $\phi: G \to D$ be a homomorphism which extends the monomorphism $G' \rightarrowtail D$ (use Theorem 19). Let $\eta: G \to \prod_{n<\omega}(G/nG)$ be defined by $\eta(g) = <g + nG>$ and let $\theta: G \to \prod_{n<\omega}(G/nG) + D$ be defined by $\theta(g) = \eta(g) + \phi(g)$. If $\theta(g) = 0$, then $\eta(g) = 0$, which implies that $g + nG = 0 + nG$ for each $n \neq 0$, that is, $g \varepsilon \bigcap_{n<\omega} nG$. But also $\phi(g) = 0$ and $g \varepsilon \bigcap_{n<\omega} nG$ further implies that $g = 0$, since ϕ is monic when restricted to $\bigcap_{n<\omega} nG$. If n divides $\theta(g)$ in $\prod_{n<\omega} G/nG + D$, then each coordinate of $\theta(g)$ is divisible by n. Hence, $g + nG = nx + nG = 0 + nG$, that is, $g \varepsilon nG$ and g is divisible by n in G. Thus, $\mathrm{Image}\theta \simeq G$ is a pure subgroup of $\prod_{n<\omega} G/nG + D$. With the aid of Theorem 12, it is an easy exercise to show that any bounded group (e.g., G/nG) is a direct summand of a direct product of torsion cyclic groups. It follows that $\prod_{n<\omega} G/nG$ is a direct summand of a direct product A of torsion cyclic groups and, therefore, $G \simeq \mathrm{Image}\theta$ is isomorphic with a pure subgroup of $A + D$.

<u>Definition</u>. An exact sequence $A_1 \xrightarrow{\phi_1} A_2 \xrightarrow{\phi_2} A_3 \xrightarrow{\phi_3} \ldots \xrightarrow{\phi_{n-1}} A_n \xrightarrow{\phi_n} A_{n+1} \longrightarrow \ldots$ is called <u>pure exact</u> if the image of ϕ_i is a pure subgroup of A_{i+1} for each i.

For the notions of a pullback and pushout of a short exact sequence and the Baer sum of two short exact sequences, see the chapter on Preliminary Facts.

<u>Lemma 26</u>. (1) A pushout of a short pure exact sequence is pure exact.

(2) A pullback of a short pure exact sequence is pure exact.

(3) The Baer sum of two short pure exact sequences is pure exact.

<u>Proof</u>. (1) Suppose that $A \overset{i}{\rightarrowtail} B \overset{\nu}{\twoheadrightarrow} C$ is pure exact and that the diagram

$$\begin{array}{ccccc} A & \overset{i}{\rightarrowtail} & B & \overset{\nu}{\twoheadrightarrow} & C \\ {\scriptstyle f}\downarrow & & \downarrow{\scriptstyle \psi} & & \| \\ E & \overset{\pi}{\rightarrowtail} & M & \overset{\rho}{\twoheadrightarrow} & C \end{array}$$

represents a pushout of $A \overset{i}{\rightarrowtail} B \overset{\nu}{\twoheadrightarrow} C$. Recall that $M = (E + B)/N$ where $N = \{(f(a),-i(a)) : a \in A\}$, $\pi : e \to (e,0) + N$, $\psi : b \to (0,b) + N$ and $\rho : (e,b) + N \to \nu(b)$. Suppose that $n(e,b) + N = \pi(e_1) = (e_1,0) + N$. Then $n(e,b) = (e_1,0) + (f(a),-i(a))$, $a \in A$, which implies that $-nb = i(a)$ and, hence, that $i(a) = na_1$, $a_1 \in A$, since $i(A)$ is pure in B. Therefore, $e_1 = ne - f(a) = ne - f(na_1) = n(e - f(a_1)) \in nE$ and so $\pi(E)$ is a pure subgroup of M.

(2) For $A \overset{i}{\rightarrowtail} B \overset{\nu}{\twoheadrightarrow} C$ pure exact, suppose that the commutative diagram

$$\begin{array}{ccccc} A & \overset{\pi}{\rightarrowtail} & M & \overset{\rho}{\twoheadrightarrow} & E \\ \| & & {\scriptstyle \psi}\downarrow & & \downarrow{\scriptstyle f} \\ A & \overset{i}{\rightarrowtail} & B & \overset{\nu}{\twoheadrightarrow} & C \end{array}$$

represents a pullback of $A \overset{i}{\rightarrowtail} B \overset{\rho}{\twoheadrightarrow} C$. Recall that $M = \{(b,e) \in B + E : \nu(b) = f(e)\}$, $\pi : a \to (a,0)$, $\rho : (b,e) \to e$ and $\psi : (b,e) \to b$. If $n(b,e) = \pi(a) = (a,0)$, then $nb = a$ and so $a = na_1 \in nA$ since A is a pure subgroup of B. The proof of (3) is similar to (1) and (2) and is left as an exercise to the reader.

The following two theorems are the analogue of Theorems 18 and 19, respectively.

<u>Theorem 27</u>. Let G be a group. Then the following are equivalent:

(1) G is pure projective.

(2) Every pure short exact sequence $A \rightarrowtail B \twoheadrightarrow G$ splits.

(3) G is a direct sum of cyclic groups.

Proof. The proofs (1) $\Longrightarrow$ (2) and (2) $\Longrightarrow$ (3) are left as exercises. To prove (3) $\Longrightarrow$ (1) we use a different style than in the proof of Theorem 18.

(3) $\Longrightarrow$ (1). Suppose that the epimorphism $\nu: B \twoheadrightarrow C$ has a pure kernel and that $f: G \to C$ is a homomorphism. We now form the pullback of the pure exact sequence $\operatorname{Ker}\nu \rightarrowtail B \overset{\nu}{\twoheadrightarrow} C$.

$$\begin{array}{ccccc} \operatorname{Ker}\nu & \rightarrowtail & M & \overset{\rho}{\twoheadrightarrow} & G \\ \| & & \downarrow \psi & & \downarrow f \\ \operatorname{Ker}\nu & \rightarrowtail & B & \overset{\nu}{\twoheadrightarrow} & C \end{array}$$

By Lemma 26(2), the top row is pure exact and, therefore, by Lemma 7(b), the top row splits. Hence, there is a homomorphism $\theta: G \to M$ such that $\rho\theta = 1_G$. With $\phi = \psi\theta: G \to B$, we have that $\nu\phi = \nu\psi\theta = f\rho\theta = f1_G = f$. Thus, G is pure projective.

Theorem 28. Let G be a group. Then the following are equivalent:

(1) G is pure injective.

(2) Every pure short exact sequence $G \rightarrowtail B \twoheadrightarrow C$ splits.

(3) $G = H + D$ where H is a direct summand of a direct product of cyclic torsion groups and D is divisible.

Proof. Again we leave (1) $\Longrightarrow$ (2) and (2) $\Longrightarrow$ (3) as exercises and concentrate our attention on (3) $\Longrightarrow$ (1): Suppose that G satisfies the hypothesis of (3), suppose that $i: A \rightarrowtail B$ has a pure image in B and suppose that $f: A \to G$ is a homomorphism. We construct the pushout diagram

$$\begin{array}{ccccc} A & \overset{i}{\rightarrowtail} & B & \overset{\nu}{\twoheadrightarrow} & B/A \\ f \downarrow & & \downarrow \psi & & \| \\ G & \overset{\pi}{\rightarrowtail} & M & \overset{\rho}{\twoheadrightarrow} & B/A \end{array}$$

By Lemma 26(1), the bottom row is pure exact and by Exercises 1 and 2 the bottom row splits. Hence, there is a homomorphism $\theta: M \to G$ such that $\theta\pi = 1_G$. With $\phi = \theta\psi: B \to G$, we have that $\phi i = \theta\psi i = \theta\pi f = 1_G f = f$. Thus,

G is pure injective.

We remark here that the notion of pure injectivity above coincides with Kaplansky's notion of algebraic compactness (see Infinite Abelian Groups, 69). In fact, we shall often use this terminology where it is traditional to do so. In the discussion that follows, we shall see that the concept of algebraic compactness can be derived in an entirely different fashion.

Definition. Let G be a group. We define the n-adic (or Z-adic) topology on G by taking the collection $[nG]_{n<\omega}$ $(n \neq 0)$ as a basis of neighborhoods of zero.

We observe that G is a topological group in the n-adic topology (that is, the taking of inverses is a continuous function and addition is also continuous) and that G is Hausdorff if and only if $\bigcap_{n<\omega} nG = 0$. Furthermore, a sequence $\{g_n\}_n$ in G converges to g if and only if $g_n - g = r_n! x_n \in G$ where $r_n \to \infty$ as $n \to \infty$.

Definition. If G is Hausdorff in the n-adic topology, we define the completion $\hat{G}$ of G to be the group of Cauchy sequences modulo null sequences (that is, sequences which converge to zero). We remark that G is cannonically embedded in $\hat{G}$ by identifying G with the equivalence classes of constant sequences in $\hat{G}$. Finally, whenever we say in this section that a group is Hausdorff or complete, we always mean with respect to the n-adic topology.

Lemma 29. Let G be a group that is Hausdorff in its n-adic topology. Then $\hat{G}$ is again Hausdorff and complete in the n-adic topology on $\hat{G}$.

Proof. Suppose that $z \in \bigcap_{n<\omega} n\hat{G}$. Then $z = ny$ where $y \in n\hat{G}$. Let $\{z_i\}_i$ and $\{y_i\}_i$ be sequences in G converging to z and y, respectively. Then $z_i - ny_i \to 0$. Hence, for sufficiently large i, we have that $z_i - ny_i$ is divisible by n and hence so is z_i. Since this is true for any $n > 0$, it follows that $z_i \to 0$ in G. Therefore, $z = 0$.

If $\{x_i\}_i$ is a Cauchy sequence in G, we choose y_i in $\hat{G}$ "very close" to

x_i so that $\{y_i\}_i$ is a Cauchy sequence in G. Then $y_i \to x$ in $\hat{G}$ and, therefore, it is easy to see that $x_i \to x$ in $\hat{G}$. Thus, $\hat{G}$ is complete.

We note that, if A is a pure subgroup of B, then the n-adic topology on A coincides with the relative topology on A induced by the n-adic topology on B.

Lemma 30. Let G be Hausdorff in its n-adic topology. Then G is a pure subgroup of its n-adic completion $\hat{G}$ and $\hat{G}/G$ is divisible.

Proof. Assume that $g \in G$ and that n divides g in $\hat{G}$. Suppose that $nx = g$, $x \in \hat{G}$, and that the sequence $\{y_i\}_i$ in G converges to x. Then $g - ny_i \to 0$ which implies that $g - ny_i \in nG$ for large i. Thus, $g \in nG$ and G is pure in $\hat{G}$. One easily sees (Exercise 10) that G is dense in the n-adic topology on $\hat{G}$ if and only if $\hat{G}/G$ is divisible and, by the construction of $\hat{G}$, G is certainly dense in the n-adic topology on G.

Lemma 31. If G is a pure subgroup of a group H which is Hausdorff and complete in the n-adic topology, then the closure $\overline{G}$ of G in the n-adic topology on H is a pure subgroup of H and $\overline{G} \simeq G$.

Proof. Extending the identity map on G to an isomorphism ϕ between $\hat{G}$ and $\overline{G}$ is accomplished by $\phi(\lim g_n) = \lim g_n$ where $\{g_n\}_n$ is a Cauchy sequence in G. Hence, $\overline{G}/G \simeq \hat{G}/G$ is divisible which shows that $\overline{G}/G$ is pure in H/G. Thus, by Lemma 5(ii), $\overline{G}$ is necessarily pure in H.

Definition. Recall that $I_p = \{n/m \in Q\text{: p and m are relatively prime}\}$, that is, I_p is the localization of the ring of integers to the prime p. If a group G is uniquely divisible by each positive integer relatively prime to p, then one can clearly define $r \cdot g \in G$ for each $r \in I_p$ and $g \in G$. In this case we call G an I_p-module. For example, any p-group is naturally an I_p-module. Furthermore, the topology on such a group G defined by taking the subgroups $[p^nG]_{n<\omega}$ as neighborhoods of zero coincides with the n-adic topology on G. The topology on G with $[p^nG]_{n<\omega}$ as neighborhoods of zero is referred to the p-adic topology on G.

Theorem 32. Let G be a group which is naturally an I_p-module and which is Hausdorff in its p-adic topology. Then $\hat{G}$ is also an I_p-module which is Hausdorff and complete in its p-adic topology. Moreover, $\hat{G}$ is algebraically compact.

Proof. Since the n-adic and p-adic topologies coincide on G, the n-adic and p-adic completions of G coincide. Hence, the first part of the theorem is clear.

Suppose that G is an I_p-module which is Hausdorff and complete in the p-adic topology. Since $\bigcap_{n<\omega} p^n G = \bigcap_{n<\omega} nG = 0$ and since $nG = p^m G$, where $n = p^m q$ and p and q are relatively prime, then the pure embedding described in the proof of Theorem 25 shows that G is a pure subgroup of a group A which is a direct product of cyclic p-groups. We remark that A is both algebraically compact (Theorem 28) and an I_p-module. We choose a subgroup H of A containing G such that H/G is a p-basic subgroup of A/G. By Lemma 7(a), H = G + B where B is a direct sum of cyclic groups. By Lemma 5(iii), H is p-pure in A. Since A is uniquely m divisible for m and p relatively prime, we define $B_* = \{x \neq 0 \in A: mx \in B \text{ where } m \text{ and } p \text{ are relatively prime}\}$. Since $(G + B_*)/H \simeq B_*/B$ has zero p-torsion, it follows that $G + B_*$ is p-pure in G. Moreover, $G + B_*$ is q-divisible for each prime $q \neq p$ since both G and B_* have this property. By Exercise 4 (Chapter II), we have that $G + B_*$ is pure and, by choice of H, that $A/(G + B_*)$ is divisible. By Lemma 31, $A = \overline{G + B_*} \simeq \widehat{G + B_*} \simeq \hat{G} + \hat{B}_* \simeq G + \hat{B}_*$ since $G = \hat{G}$. Hence, G is isomorphic to a direct summand of A and thus Theorem 28 yields that G is pure injective.

Our next theorem provides a remarkable connection between the notion of algebraic compactness and the notion of completeness in the n-adic topology.

Theorem 33. (Maranda) A group G, which is Hausdorff in the n-adic topology, is algebraically compact if and only if G is complete in its n-adic topology.

Proof. First suppose that G is Hausdorff and that G is algebraically compact. Then Theorem 28 shows that G is a direct summand of a direct product A of cyclic torsion groups. It is elementary that A is complete in its n-adic topology and, hence, that a direct summand, namely G, is also complete.

Now suppose that G is Hausdorff and complete in its n-adic topology. By Theorem 25, we may assume that G is a pure subgroup of a group $A = \prod_{primes} A^p$ where A^p is a direct product of cyclic p-groups for each prime p. Since G is pure in A and G is complete, then G is necessarily a closed subgroup of A. Let G^p denote the coordinate projection of G into A^p and let $x_p \in G^p$. Hence, there is $x \in G$ whose p-th coordinate is x_p. For each positive integer n, we write $n! = p^{e_n} r_n$, where p and r_n are relatively prime. Therefore, there are integers s_n and t_n such that $1 = s_n p^{e_n} + t_n r_n$. Define $g_n = t_n r_n x \in G$. Let $y = x - x_p \in \prod_{q \neq p} A^q$. Then $x_p - g_n = (1 - t_n r_n) x_p - t_n r_n y = s_n p^{e_n} x_p - t_n r_n y$. Since $n! A^p = p^{e_n} r_n A^p = p^{e_n} A^p$ and $n!(\prod_{q \neq p} A^q) = r_n (p^{e_n} \prod_{q \neq p} A^p) = r_n (\prod_{q \neq p} A^q)$, it follows that $s_n p^{e_n} x_p \in n! A^p$ and $t_n r_n y \in n!(\prod_{q \neq p} A^p)$ and thus that $x_p - g_n \in n!A$. Therefore, $g_n \to x_p$ and so $x_p \in G$, since G is closed in A. We now have that $\sum_p G^p \subseteq G \subseteq \prod_p G^p \subseteq \prod_p A^p = A$. By Exercise 9, $\sum_p G^p$ is dense in the n-adic topology on $\prod_p G^p$ which implies that $G = \prod G^p$. Furthermore, for each prime p, $G^p = G \cap A^p$ is pure and closed in A^p. By Theorem 32, G^p is a direct summand of A^p for each prime p. Hence, G is a direct summand of A which shows, by Theorem 28, that A is algebraically compact.

Further examination of the above proof yields ...

<u>Corollary 34</u>. A reduced algebraically compact group G has the form, $G = \prod_{primes} G^p$ where G^p is an I_p-module that is Hausdorff and complete in its p-adic topology.

<u>Corollary 35</u>. I_p^* (p-adic group) is algebraically compact for each prime p.

The following corollary is left as an exercise (see Exercise 11).

<u>Corollary 36</u>. For any torsion group T and any group X, Hom(T,X) is

algebraically compact as an abelian group.

Exercises

1. Suppose that each A_i, $i \in I$, is a direct summand of a group whenever it is a pure subgroup (from Lemma 6, torsion cyclic groups have this property). Prove that $A = \prod_{i \in I} A_i$ has the same property.

2. If A has the property described in Exercise 1, then so does any direct summand.

3. Prove the remaining parts of Theorem 27.

4. Prove the remaining parts of Theorem 28.

5. In Theorems 18 and 19, prove (3) $\Longrightarrow$ (1) in the style of Theorems 27 and 28, respectively.

6. Prove part (3) in Lemma 26, that is, the Baer sum of two short pure exact sequences is again pure exact.

7. Using the definition of Ext(C,A) as an abelian group given in Chapter 0, using Lemma 26 and Exercise 6, prove that the equivalence classes corresponding to pure exact sequences $A \rightarrowtail M \twoheadrightarrow C$ form a subgroup of Ext(C,A) (usually denoted by Pext(C,A)).

8. If A^p is an I_p-module for each prime p, show that the n-adic topology on $\prod_p A^p$ coincides with the product of their respective p-adic topologies.

9. If A^p is an I_p-module for each prime p, show that $\sum_p A^p$ is dense in $\prod_p A^p$ in the n-adic topology.

10. Let A be a subgroup of B. Prove that A is dense in the n-adic topology on B if and only if B/A is divisible.

11. Prove that Hom(T,X) is algebraically compact for any torsion group T and any group X using Theorem 33.

12. Show that $\hat{Z} = \prod_p I_p^*$.

13. Suppose that a p-group G is a direct summand of every p-group in

which it is a pure subgroup. Prove that $G = tA$ for some algebraically compact group A. Such a group is called <u>torsion complete</u>.

14. Let $B = \sum_{n<\omega} Z(p^n)$ and let $\overline{B} = t\hat{B}$, i.e., $\overline{B}$ is the torsion completion of B. Prove

(i) $\overline{B}$ is a p-group.

(ii) $|\overline{B}| = 2^{\aleph_0}$.

(iii) If C is a direct summand of $\overline{B}$ and if C is a direct sum of cyclic groups, then C is bounded.

(iv) B is a basic subgroup of $\overline{B}$.

15. Let G be a p-group and show that $\infty > h^G(x + y) > h^G(x) + h^G(y)$ can occur.

16. Find an example of a group G and pure subgroups A and B with $A \cap B = 0$ such that A + B is not pure in G.

17. If A_i is essential in B_i, show that $\sum A_i$ is an essential subgroup of $\sum B_i$.

18. If $X = [x_i]_{i\in I}$ is a maximal independent subset of a group G such that the order of each x_i is infinite or of prime power, then $|X| = \text{rank}(G)$. If G is uncountable, show that $|X| = |G| = \text{rank}(G)$.

19. If G is Hausdorff and torsion free, prove that $\hat{G}$ is also torsion free.

20. Prove that a reduced torsion group is algebraically compact if and only if it is bounded.

21. A subgroup A of a group B is called a <u>pure essential subgroup</u> if the only subgroup C of B with the properties that $A \cap C = 0$ and A + C pure in B is the zero subgroup. Prove that every group G can be embedded as a pure essential subgroup of a pure injective group A. (<u>Hint</u>: Exhibit pure and dense embedding of G into $\widehat{(G/G')} + D(G')$, where $G' = \bigcap_{n<\omega} nG$.)

22. Prove the uniqueness of the embedding in Exercise 21 in the sense of Theorem 20.

IV. STRUCTURE OF THE TENSOR PRODUCT AND THE GROUP OF EXTENSIONS

In this chapter we begin to make use of the homological concepts discussed in Chapter 0. In particular, for definitions of Hom, Ext, $\otimes$ and Tor and their exact sequences, see Chapter 0, Preliminary Facts.

Proposition 37. If $A \rightarrowtail B \twoheadrightarrow C$ is a pure exact sequence and if X is any group, then $X \otimes A \rightarrowtail X \otimes B \twoheadrightarrow X \otimes C$ is exact.

Proof. It suffices to show that the connecting homomorphism ∂_X: $\mathrm{Tor}(X,C) \to X \otimes A$ is the zero map. Let $\langle x,m,c\rangle \in \mathrm{Tor}(X,C)$. Therefore, $mx = mc = 0$ and, by definition, $\partial_X\langle x,m,c\rangle = x \otimes a$ where $c = b + A$ and $mb = a$. But $mb = ma'$ since A is pure and so $x \otimes a = x \otimes ma' = mx \otimes a' = 0 \otimes a' = 0$. It follows that $\partial_X = 0$.

The following result shows that the tensor product has a simplifying effect on the class of torsion groups.

Theorem 38. If G and H are torsion groups, then $G \otimes H$ is a direct sum of cyclic groups.

Proof. Since $\otimes$ commutes with direct sums and since $G_p \otimes H_q = 0$ for $p \neq q$ (see Exercise 3), we may assume that G and H are both p-groups for a fixed prime p. Let B be a basic subgroup of G. By Proposition 37, there is an exact sequence $B \otimes H \rightarrowtail G \otimes H \twoheadrightarrow (G/B) \otimes H$ and $(G/B) \otimes H = 0$ by Exercise 2. So $B \otimes H \approx G \otimes H$. For B' a basic subgroup of H, a similar argument shows that $B \otimes H \approx B \otimes B'$, that is, $B \otimes B' \approx G \otimes H$. But $B \otimes B'$ is a direct sum of cyclic groups since $\otimes$ commutes with direct sums.

Our final result in this chapter on tensor products reveals to some extent the role of the commutative diagram in determining information.

Theorem 39. If A and B are groups, then $t(A \otimes B) = \{tA \otimes B, A \otimes tB\}$ and $A \otimes B/t(A \otimes B) \approx (A/tA) \otimes (B/tB)$.

Proof. The pure exact sequences $tA \rightarrowtail A \twoheadrightarrow A/tA$ and

$tB \rightarrowtail B \twoheadrightarrow B/tB$ induce (using Proposition 37) the commutative diagram

$$\begin{array}{ccccc}
tA \otimes tB & \rightarrowtail & tA \otimes B & \overset{\eta}{\twoheadrightarrow} & tA \otimes (B/tB) \\
\downarrow & & \downarrow & & \downarrow \pi \\
A \otimes tB & \rightarrowtail & A \otimes B & \overset{\phi}{\twoheadrightarrow} & A \otimes (B/tB) \\
\downarrow & & \downarrow \theta & & \downarrow g \\
(A/tA) \otimes tB & \rightarrowtail & (A/tA) \otimes B & \overset{f}{\twoheadrightarrow} & (A/tA) \otimes (B/tB)
\end{array}$$

Let $\nu = f\theta = g\phi$. It follows that $\{A \otimes tB, tA \otimes B\} = \{\mathrm{Ker}\phi, \mathrm{Ker}\theta\} \subseteq \mathrm{Ker}\nu$. Now suppose that $\nu(x) = 0$. Therefore, $\phi(x) \in \mathrm{Ker}g = \mathrm{Image}\pi$ and so $\phi(x) = \pi(y)$. But $y \in \mathrm{Image}\eta$ implies that $y = \eta(x_1)$, where $x_1 \in tA \otimes B = \mathrm{Ker}\theta$. Moreover, $\phi(x - x_1) = \phi(x) - \phi(x_1) = \phi(x) - \pi\eta(x_1) = \phi(x) - \pi(y) = \phi(x) - \phi(x) = 0$. Hence, $x - x_1 = x_2 \in \mathrm{Ker}\phi$ and so $x \in \{\mathrm{Ker}\phi, \mathrm{Ker}\theta\}$. It follows that $A \otimes B/\{A \otimes tB, tA \otimes B\} \simeq (A/tA) \otimes (B/tB)$. Since $\{A \otimes tB, tA \otimes B\}$ is torsion and since by Exercise 8 $(A/tA) \otimes (B/tB)$ is torsion free, the theorem is proved.

We now switch our attention to the structure and properties of Ext(A,B).

Definition. After Harrison [47], a group E is called cotorsion if Ext(X,E) = 0 for all torsion free groups X. It is elementary to see that a group T is torsion if and only if Hom(T,X) = 0 for all torsion free groups X. It is in this sense that torsion and cotorsion are dual concepts.

Lemma 40. A group E is cotorsion if and only if Ext(Q,E) = 0.

Proof. Since the necessity is clear, we suppose that Ext(Q,E) = 0. Therefore, $\mathrm{Ext}(\sum_\alpha Q, E) \simeq \prod_\alpha \mathrm{Ext}(Q,E) = 0$. It follows that, for any torsion free group A, Ext(D(A),E) = 0. This fact together with the exactness of the sequence $\mathrm{Ext}(D(A)/A,E) \rightarrow \mathrm{Ext}(D(A),E) \twoheadrightarrow \mathrm{Ext}(A,E)$ implies that Ext(A,E) = 0 and thus E is cotorsion.

Lemma 41. If D is divisible, then Hom(A,D) is algebraically compact. In particuiar, $\mathrm{Hom}(A,D) \simeq \mathrm{Hom}(A/tA,D) + \mathrm{Hom}(tA,D)$, where Hom(A/tA,D) is divisible and Hom(tA,D) is reduced and algebraically compact.

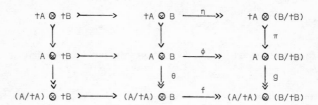

Proof. Let B = A/tA. Then D(B) is torsion free and divisible and, furthermore, it is elementary to observe that $\mathrm{Hom}(D(B),D) \approx \Pi\mathrm{Hom}(Q,D)$ is divisible. The exactness of the sequence

$\mathrm{Hom}(D(B),D) \longrightarrow \mathrm{Hom}(B,D) \xrightarrow{\delta} \mathrm{Ext}(D(B)/B,D) = 0$ (recall that D is injective) shows that $\mathrm{Hom}(B,D) = \mathrm{Hom}(A/tA,D)$ is the homomorphic image of a divisible group and, hence, is itself divisible. Consider the exact sequence $\mathrm{Hom}(A/tA,D) \rightarrowtail \mathrm{Hom}(A,D) \longrightarrow \mathrm{Hom}(tA,D) \longrightarrow \mathrm{Ext}(A/tA,D) = 0$ (again recall that D is injective). The divisibility of $\mathrm{Hom}(A/tA,D)$ now yields that $\mathrm{Hom}(A,D) \approx \mathrm{Hom}(A/tA,D) + \mathrm{Hom}(tA,D)$. By Corollary 36, $\mathrm{Hom}(tA,D)$ is reduced and algebraically compact.

A connection between algebraically compact and cotorsion is contained in ...

Theorem 42. A homomorphic image of an algebraically compact group is cotorsion.

Proof. Suppose that A is algebraically compact and that E is a homomorphic image of A. This yields an epimorphism $\mathrm{Ext}(Q,A) \twoheadrightarrow \mathrm{Ext}(Q,E)$. Hence, it suffices to establish that $\mathrm{Ext}(Q,A) = 0$. But every extension of A by Q is necessarily pure and so splits since A is pure injective. Thus, $\mathrm{Ext}(Q,A) = 0$.

Corollary 43. $\mathrm{Ext}(A,B)$ is cotorsion for any groups A and B.

Proof. The exact sequence $B \rightarrowtail D(B) \twoheadrightarrow D(B)/B$ induces the exact sequence $\mathrm{Hom}(A,D(B)/B) \xrightarrow{\delta} \mathrm{Ext}(A,B) \longrightarrow \mathrm{Ext}(A,D(B)) = 0$. By Lemma 41, $\mathrm{Hom}(A,D(B)/B)$ is algebraically compact and, by Theorem 42, $\mathrm{Ext}(A,B)$ is cotorsion.

Proposition 44. (a) $\mathrm{Ext}(T,B)$ is a reduced cotorsion group for any torsion group T.

(b) $\mathrm{Ext}(Q,B)$ is torsion free and divisible for any group B.

Proof. (a) We again use the exact sequence $B \rightarrowtail D(B) \twoheadrightarrow D(B)/B$ to induce the exact sequence $\mathrm{Hom}(T,D(B)) \xrightarrow{\theta} \mathrm{Hom}(T,D(B)/B) \xrightarrow{\delta} \mathrm{Ext}(T,B) \longrightarrow \mathrm{Ext}(T,D(B)) = 0$. Let E be the image of $\theta = \mathrm{Ker}\,\delta$. Since $\mathrm{Hom}(T,D(B))$ is algebraically compact, we have by Theorem 42 that E is cotorsion. From the exact sequence $E \rightarrowtail \mathrm{Hom}(T,D(B)/B) \twoheadrightarrow \mathrm{Ext}(T,B)$ we obtain the exact sequence $\mathrm{Hom}(Q,\mathrm{Hom}(T,D(B)/B)) \longrightarrow \mathrm{Hom}(Q,\mathrm{Ext}(T,B)) \longrightarrow \mathrm{Ext}(Q,E)$. Since E is cotorsion, $\mathrm{Ext}(Q,E) = 0$ and $\mathrm{Hom}(Q,\mathrm{Hom}(T,D(B)/B)) = 0$ by Corollary 36. Hence, $\mathrm{Hom}(Q,\mathrm{Ext}(T,B)) = 0$ and so $\mathrm{Ext}(T,B)$ is reduced. Finally, $\mathrm{Ext}(T,B)$ is cotorsion from Corollary 43.

(b) By Exercise 13, $\mathrm{Hom}(Q,X)$ is divisible for any X. Therefore, in the exact sequence $\mathrm{Hom}(Q,D(B)) \longrightarrow \mathrm{Hom}(Q,D(B)/B) \xrightarrow{\delta} \mathrm{Ext}(Q,B) \longrightarrow \mathrm{Ext}(Q,D(B)) = 0$, the image D of $\mathrm{Hom}(Q,D(B))$ in $\mathrm{Hom}(Q,D(B)/B)$ is divisible. Hence, $\mathrm{Hom}(Q,D(B)/B) \approx D + \mathrm{Ext}(Q,B)$. Thus, it remains to show that $\mathrm{Hom}(Q,X)$ is torsion free for any X. If $f \in \mathrm{Hom}(Q,X)$ and $nf = 0$, this implies that Image f is bounded and divisible and, hence Image $f = 0$, that is, $f = 0$. This completes our proof.

The next two lemmas establish important properties of cotorsion groups.

Lemma 45. (1) A homomorphic image of a cotorsion group is cotorsion; in particular, direct summands of cotorsion groups are cotorsion.

(2) A direct product of cotorsion groups is cotorsion.

(3) If E is cotorsion and $E_o \subseteq E$ such that E/E_o is reduced, then E_o is cotorsion. If E is reduced and E/E_o is torsion free, then the previous condition is also necessary.

(4) A reduced, torsion free cotorsion group is algebraically compact.

Proof. (1) If E cotorsion and $E \twoheadrightarrow E_1$ is an epimorphism, then there is an epimorphism $0 = \mathrm{Ext}(Q,E) \twoheadrightarrow \mathrm{Ext}(Q,E_1)$ and so $\mathrm{Ext}(Q,E_1) = 0$. Thus, E_1 is cotorsion.

(2) If each E_i, $i \in I$, is cotorsion, then $\mathrm{Ext}(Q,\prod_i E_i) \approx \prod_i \mathrm{Ext}(Q,E_i) = 0$ and so $\prod_i E_i$ is cotorsion.

(3) Since E/E_o is reduced, we have $\mathrm{Hom}(Q,E/E_o) = 0$. From the exactness of $0 = \mathrm{Hom}(Q,E/E_o) \to \mathrm{Ext}(Q,E_o) \to \mathrm{Ext}(Q,E) = 0$, it follows that $\mathrm{Ext}(Q,E_o) = 0$ and that E_o is cotorsion. If E_o is cotorsion and E/E_o is torsion free, then $E \approx E_o + (E/E_o)$ since $\mathrm{Ext}(E/E_o,E_o) = 0$. Thus, if E is reduced, E/E_o is necessarily reduced.

(4) Let E be a reduced torsion free group. Then $\bigcap_{n<\omega} nE = 0$ since E is torsion free, that is, E is Hausdorff in its n-adic topology. Hence, we have a pure exact sequence $E \rightarrowtail \hat{E} \twoheadrightarrow D$ where D is divisible. By Exercise 19 (Chapter III), $\hat{E}$ is torsion free and, hence, D is necessarily torsion free. Thus, $E \rightarrowtail \hat{E} \twoheadrightarrow D$ splits if and only if $E = \hat{E}$, that is, E is cotorsion if and only if E is algebraically compact.

Lemma 46. If $A \rightarrowtail B \twoheadrightarrow C$ is exact with C torsion free and if $f: A \to E$ is a homomorphism from A into the cotorsion group E, then there is a homomorphism $\psi: B \to E$ such that the diagram

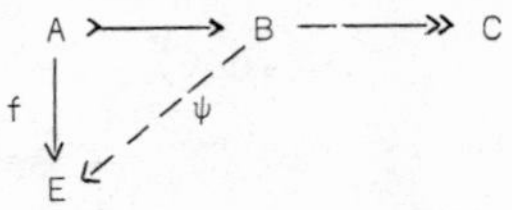

is commutative.

Proof. This result is an obvious consequence of the exact sequence $\mathrm{Hom}(B,E) \to \mathrm{Hom}(A,E) \overset{\delta}{\to} \mathrm{Ext}(C,E) = 0$.

Theorem 47. Every reduced group G can be embedded in a reduced cotorsion group E such that E/G is torsion free and divisible. Moreover, this embedding is unique in the sense that, if G is contained in a reduced cotorsion group E' with E'/G torsion free and divisible, then there is an isomorphism ψ between E and E' which preserves the identity map on G.

Proof. Let G be a reduced group. Since $\mathrm{Hom}(Q,G) = 0$ and $G \cong \mathrm{Hom}(Z,G)$, the exact sequence $Z \rightarrowtail Q \twoheadrightarrow Q/Z$ induces the exact sequence $G \cong \mathrm{Hom}(Z,G) \rightarrowtail \mathrm{Ext}(Q/Z,G) \longrightarrow \mathrm{Ext}(Q,G) \twoheadrightarrow \mathrm{Ext}(Z,G) = 0$. The last zero is due to the fact that Z is projective. By Proposition 44, we have that $E = \mathrm{Ext}(Q/Z,G)$ is a reduced cotorsion group and that $\mathrm{Ext}(Q,G)$ is torsion free and divisible.

Now suppose that G is a subgroup of a reduced cotorsion group E' such that E'/G is torsion free and divisible. Lemma 46 then guarantees a commutative diagram

$$\begin{array}{ccccc} G & \overset{i_1}{\rightarrowtail} & E & \twoheadrightarrow & E/G \\ {\scriptstyle i_2}\downarrow & \swarrow{\scriptstyle \psi} & & & \\ E' & & & & \end{array}$$

where i_1 and i_2 are inclusion maps. Let $H = \text{Image}\psi$ and $K = \text{Ker}\psi$. Lemma 45(1) yields that H is cotorsion and Lemma 45(3) implies that K is also cotorsion. The induced map $\psi^*: E/G \twoheadrightarrow H/G$ shows that H/G is divisible which implies that $E'/H \simeq E'/G/H/G$ is torsion free and divisible. By Lemma 45(3), necessarily $H = E'$. Furthermore, ψ induces an isomorphism $E/(G + K) \simeq E'/G$ which implies that $G + K$ is pure in E. Therefore, $K \simeq \{K,G\}/G$ is a pure subgroup of $E(G)/G$, that is, K is divisible. However, $K = 0$ since E is reduced. Thus, ψ is the desired isomorphism.

Definition. For a reduced group G we shall denote $\text{Ext}(Q/Z,G)$ by $E(G)$ and we refer to $E(G)$ as the cotorsion completion of G where the implied uniqueness is to be taken in the sense of Theorem 47.

Definition. After Harrison [47], a reduced cotorsion group E is called adjusted if E admits no nonzero torsion free direct summand.

Harrison's notion of duality yields the following explicit relationship between the class of all reduced torsion groups and the class of all adjusted cotorsion groups.

Theorem 48. (Harrison) There is a bijective correspondence between the class of all reduced torsion groups and the class of all adjusted cotorsion groups, namely, the one $T \to E(T) = \text{Ext}(Q/Z,T)$.

Proof. The fact that $E(T)$ is adjusted, for a reduced torsion group T, follows easily from the recollection that $E(T)$ is reduced while $E(T)/T$ is torsion free and divisible. Furthermore, Theorem 47 shows that this correspondence is one-to-one. It remains only to prove that this correspondence is onto. So let E be an adjusted cotorsion group with torsion subgroup T. Then $E/T = H/T + K/T$ where H/T is torsion free and divisible and K/T is reduced and torsion free. Since $E/H \simeq K/T$, it follows by Lemma 45(3) that

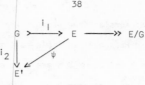

$E \simeq H + (K/T)$. But E an adjusted cotorsion group implies that $H = E$ and Theorem 47 further implies that $E = H \simeq E(T)$. Thus, $T \to E$ and so the above correspondence is onto.

<u>Lemma 49</u>. Let G be a p-group. Then G, Hom(G,X) and Ext(G,X) are naturally modules over I_p^*.

<u>Proof</u>. We already have that G is an I_p-module. Since $I_p^* = \hat{I}_p$, for $\pi \in I_p^*$, there is a sequence $\{\pi_n\}_n$ in I_p such that $\pi_n \to \pi$ in the p-adic topology on I_p^*. Hence, $\{\pi_n\}_n$ is a Cauchy sequence in I_p. Let $g \in G$ of order p^m. Then there is an $N > 0$ such that $\pi_n - \pi_r \in p^m I_p$ for $n,r \geq N$. Hence, $\pi_n g = \pi_r g$ for $n,m \geq N$. Thus, $\{\pi_n g\}_n$ is an eventually constant sequence in G. So we define $\pi g = \lim_n \pi_n g \in G$.

If X is any group and if $\pi \in I_p^*$ we make Hom(G,X) into a module over I_p^* by defining πf, where $f \in \text{Hom}(G,X)$, as follows: $(\pi f)(g) = f(\pi g)$.

Let $X \rightarrowtail D(X) \overset{\nu}{\twoheadrightarrow} D$ be an injective resolution of X and consider the exact sequence $\text{Hom}(G,D(X)) \overset{\nu_*}{\longrightarrow} \text{Hom}(G,D) \overset{\delta}{\twoheadrightarrow} \text{Ext}(G,X)$. Let $K = \text{Image}\nu_* \subseteq \text{Hom}(G,D)$. Then $K = \{f \in \text{Hom}(G,D)\colon f = \nu\psi$, where $\psi \in \text{Hom}(G,D(X))\}$. If $\pi \in I_p^*$ and $f \in K$, say $f = \nu\psi$ where $\psi \in \text{Hom}(G,D(X))$, then it is easy to observe that $\pi f = \pi\nu\psi = \nu(\pi\psi)$. Hence, $\pi f \in K$ and thus K is naturally an I_p^*-submodule of Hom(G,D). It follows that $\text{Ext}(G,X) \simeq \text{Hom}(G,D)/K$ is naturally a module over I_p^*. The details that the above definitions actually make G and Hom(G,X) into I_p^*-modules are left as an exercise (Exercise 17).

<u>Definition</u>. If G is naturally (as above) an I_p^*-module, we simply call G a p-adic module.

We are now in a position to give a general description of the structure of reduced cotorsion groups. This result generalizes our structural result (Corollary 34) for reduced algebraically compact groups since, of course, algebraically compact groups are cotorsion.

<u>Theorem 50</u>. If E is a reduced cotorsion group, then $E = \prod_p E^p$, where E^p is a p-adic module for each prime p. Moreover, $E^p = A^p + M^p$ where A^p

is adjusted and M^p is a torsion free group that is Hausdorff and complete in its p-adic topology.

Proof. From the exact sequence $Z \rightarrowtail Q \twoheadrightarrow Q/Z$, we obtain the induced exact sequence $0 = \mathrm{Hom}(Q,E) \to \mathrm{Hom}(Z,E) \to \mathrm{Ext}(Q/Z,E) \to \mathrm{Ext}(Q,E) = 0$. Hence, $E \simeq \mathrm{Hom}(Z,E) \simeq \mathrm{Ext}(Q/Z,E) \simeq \prod_p \mathrm{Ext}(Z(p^\infty),E)$. We let $E^p = \mathrm{Ext}(Z(p^\infty),E)$. By Lemma 49, E^p is a p-adic module and E^p is reduced since E is reduced. This implies that any torsion free direct summand of E^p is Hausdorff in the p-adic topology. Let $T_p = tE^p$. Then $E^p/T_p = (A^p/T_p) + (H/T_p)$, where A^p/T_p is torsion free and divisible while H/T_p is torsion free and reduced. By Lemma 45(3), we have that $E^p = A^p + M^p$. Since M^p is necessarily torsion free, a Hausdorff I_p-module and cotorsion, it follows by Lemma 45(4) and Corollary 34 that M^p is complete in the p-adic topology. Clearly, $A^p = E(T_p)$, which completes the proof.

We have seen that cotorsion groups arise as homomorphic images of algebraically compact groups and also as groups of extensions. However, they arise in another remarkable fashion as was shown by Balcerzyk [3]. In fact, it might be stated that the importance of algebraically compact and cotorsion groups is greatly enhanced by the fact that they seem to naturally appear on the scene in a variety of different ways. The following proof is modeled after one given by Nunke [107].

Theorem 51. (Balcerzyk) Let $[C_n]_{n<\omega}$ be any countable family of groups. Then $A = \prod_{n<\omega} C_n / \sum_{n<\omega} C_n$ is algebraically compact.

Proof. First we suppose that C_n is free for each n and let $S = \sum_{n<\omega} C_n$ and $P = \prod_{n<\omega} C_n$. Let π be the natural projection of P onto $P/S = A$. Since the sequence $\mathrm{Ext}(Q,P) \xrightarrow{\pi_*} \mathrm{Ext}(Q,A)$ is exact, in order to establish that A is cotorsion, it is enough to show that $\pi_* = 0$. In terms of extensions this means that for each extension $P \rightarrowtail E \overset{\nu}{\twoheadrightarrow} Q$ there is a homomorphism $f\colon E \to A$ which extends π.

For $n < \omega$, let e^n be in E such that $\nu(e^n) = 1/n!$. Then E is generated by P and $[e^n]_{n<\omega}$ with relations $e^n = (n+1)e^{n+1} + a^n$, where $a^n \in P$ for each

n. If $f: \{e^1, e^2, \ldots\} \to A$ satisfies the relations $f(e^n) = (n+1)f(e^{n+1}) + \pi(a^n)$ for all n, then f can be extended to a homomorphism of E into A with the desired properties. We, therefore, want to define f on the e^n's so as to satisfy the above relations.

Since each a^n is in P, there are elements $b^n \in S$ such that $(a^n + b^n)_i = 0$ (= i-th coordinate of $a^n + b^n$) for $i < n$. We set $y^n = \sum_{k \geq n} k!(a^k + b^k)$. This sum has meaning because it is finite on each coordinate. We further define $x^n = (1/n!)y^n = \sum_{k \geq n} (k!/n!)(a^k + b^k)$. Then (*), $y^n = y^{n+1} + n!(a^n + b^n)$ so that $x^n = (n+1)x^{n+1} + a^n + b^n$. If we define $f(e^n) = \pi(x^n)$, we see that (*) implies that $f(e^n) = f(e^{n+1}) + \pi(a^n)$ because the b^n's are in S. Thus, the required homomorphism exists.

In the general case, choose for each $n < \omega$ a free group F_n and an epimorphism $F_n \twoheadrightarrow C_n$. These epimorphisms induce an epimorphism of $\prod_{n<\omega} F_n / \sum_{n<\omega} F_n \twoheadrightarrow \prod_{n<\omega} C_n / \sum_{n<\omega} C_n = A$. From the above, it follows that A is the homomorphic image of a cotorsion group and thus, by Lemma 45(1), A is a cotorsion group. By Exercise 19, $\overline{S}$ the closure of $S = \sum C_n$ in the n-adic topology on $P = \prod C_n$ is a pure subgroup of P, that is, the subgroup of elements of infinite height in P/S is pure. Hence, P/S = H + D where H is Hausdorff and D is divisible. By Lemma 45(4), H is algebraically compact. This completes our proof.

<u>Corollary 52</u>. $\prod_\alpha Z$ is not free for any $\alpha \geq \aleph_0$.

<u>Proof</u>. It suffices to show that $\prod_{\aleph_0} Z$ is not free. From above $\prod_{\aleph_0} Z / \sum_{\aleph_0} Z$ is an algebraically compact group. However, if F is a free group of cardinality $2^{\aleph_0}$ and if S is a countable subgroup of F, then clearly F/S contains a direct summand isomorphic to Z. But Z is not algebraically compact. Thus, $\prod_{\aleph_0} Z$ cannot be free.

Actually, Nunke [107] has shown that every homomorphic image of $\prod_{\aleph_0} Z$ is of the form $\prod_\alpha Z + E + D$ where $\alpha \leq \aleph_0$, E is cotorsion and D is divisible (see Theorem 153).

Our final result in this chapter is a slight generalization of

Theorem 51.

Theorem 53. Let $\aleph$ be an arbitrary cardinal number and let β be the first ordinal having cardinality $\aleph$. If $[C_\alpha]_{\alpha<\beta}$ is any family of $\aleph$ groups and if K is the subgroup of countable sequences in $\prod_{\alpha<\beta} C_\alpha$ (i.e., all vectors in ΠC_α which are zero for all but a countable number of coordinates), then $A = K/\sum_{\alpha<\beta} C_\alpha$ is algebraically compact.

Proof. Let π be the natural map of K onto A. As in the proof of Theorem 51, we first show that A is cotorsion by showing that $\pi_* = 0$, where $\pi_*: \mathrm{Ext}(Q,K) \to \mathrm{Ext}(Q,A)$. Therefore, let $K \overset{i}{\rightarrowtail} M \overset{\nu}{\twoheadrightarrow} Q$ (where i is the inclusion map) represent an element in Ext(Q,K). Since Q is countable, $M = \{K,B\}$ for some countable subgroup B of M. Hence, $K \cap B$ is a countable subgroup of K. From the definition of K, it is clear that $K = P + H$ where $K \cap B \subseteq P$ and P is a countable product of groups. Let $N = \{P,B\}$. Then $M = \{N,H\}$. If $h = z + b \in H \cap N$, where $z \in P$ and $b \in B$, then $b \in K \cap B \subseteq P$. But this implies that $h \in P \cap H = 0$ and, hence, that $N \cap H = 0$. Thus, $M = N + H$ and $K \cap N = P$. Moreover, $\sum_{\alpha<\beta} C_\alpha = S + F$, where S is the finite sequences in P and F is the group of finite sequences in H. The exactness of $P \rightarrowtail N \twoheadrightarrow Q$ and Theorem 51 yields a commutative diagram

$$\begin{array}{ccccc} P & \rightarrowtail & N & \twoheadrightarrow & Q \\ {\scriptstyle \pi_1}\downarrow & \swarrow {\scriptstyle f_1} & & & \\ P/S & & & & \end{array}$$

Let π_2 be the natural map of H onto H/F. Then $\pi = \pi_1 + \pi_2$ and the diagram

$$\begin{array}{ccccc} K = P + H & \rightarrowtail & N + H = M & \twoheadrightarrow & Q \\ {\scriptstyle \pi = \pi_1 + \pi_2}\downarrow & \swarrow {\scriptstyle f_1 + \pi_2 = f} & & & \\ A = P/S + H/F & & & & \end{array}$$

is commutative. Thus, the map π_* is the zero map. A similar argument to that given in the proof of Theorem 51 further yields that A is in fact algebraically compact.

Exercises

1. Prove that $Z \otimes G \simeq G \simeq \text{Hom}(Z,G)$ for any group G.

2. If N is a subset of the primes and if E is a group that is divisible by each prime in N, then $E \otimes G$ is also divisible by each prime in N. Moreover, $E \otimes G = 0$ if G is a torsion group with $G_p = 0$ for $p \notin N$.

3. If G is a p-group and if H is a q-group for distinct primes p and q, then $G \otimes H = 0$.

4. Give an example of two groups A and B such that A and B are both reduced but $A \otimes B$ is a nonzero divisible group.

5. Suppose that B is a bounded group with $nB = 0$. For any group X, show that $n(B \otimes X) = 0$, $n\text{Hom}(X,B) = 0$, $n\text{Hom}(B,X) = 0$, $n\text{Ext}(X,B) = 0$ and $n\text{Ext}(B,X) = 0$.

6. From the definition of Tor(A,B) in Chapter 0, prove that if A and B are groups such that $tA_p = 0$ whenever $tB_p \neq 0$, then $\text{Tor}(A,B) = 0$. Therefore, $\text{Tor}(A,B) = 0$ if (i) A or B is torsion free or if (ii) A is a p-group and B is a q-group for $p \neq q$.

7. Using the exact sequences $Z \rightarrowtail Q \twoheadrightarrow Q/Z$ and $tG \rightarrowtail G \twoheadrightarrow G/tG$, show that $\text{Tor}(G,Q/Z) \simeq tG$. More precisely, show that $\text{Tor}(G,Z(p^\infty)) \simeq tG_p$.

8. If A and B are torsion free groups, then $A \otimes B$ is also torsion free. Furthermore, prove that both A and B are isomorphic to subgroups of $A \otimes B$.

9. If A and B are rank one torsion free groups (i.e., $D(A) \simeq D(B) \simeq Q$), prove that $A \otimes B$ is also a rank one torsion free group.

10. Prove that $I_p \otimes I_p^* \simeq I_p^*$ and that $I_p^* \otimes I_p^* \simeq I_p^* + \sum_{2^{\aleph_0}} Q$.

11. If A is an arbitrary group and if G is a p-group, show that $A \otimes G \simeq B_p \otimes G$ where B_p is a p-basic subgroup of A. (<u>Hint</u>: Consider the p-pure exact sequence $B_p \rightarrowtail A \twoheadrightarrow A/B_p$.)

12. If C is a p-group which is a direct sum of countable groups and if G is any group, then Tor(C,G) is a subgroup of a d.s.c. (<u>Hint</u>: Use the

exact sequence $G \rightarrowtail D(G) \twoheadrightarrow D(G)/G$.)

13. For any group X prove that both Hom(Q,X) and Hom(X,Q) are torsion free and divisible as abelian groups.

14. If $tG_p = 0$, prove that $\mathrm{Ext}(G, I_p^*) = 0$.

15. Suppose that G is a torsion free group and suppose that $B_p \simeq \sum_m Z$ is a p-basic subgroup of G. Let ϕ be the map that identifies B_p as a p-pure subgroup in $\sum_m I_p^* \subseteq \prod_m I_p^*$. Use Exercise 14 to show that ϕ extends to a homomorphism $\theta: G \to \prod_m I_p^*$ and that the Imageθ is p-pure in $\prod_m I_p^*$.

16. Let E be a reduced torsion free, cotorsion group such that E is naturally a p-adic module. Use Exercise 15 to prove that E is a direct summand of a direct product of copies of I_p^*.

17. Carry out the details in Lemma 49.

18. Show that $E(Z) = \prod_p I_p^*$. Moreover, if C is a reduced countable torsion free group, then prove that E(C) is isomorphic to a direct summand of $E(\sum_{\aleph_0} Z)$.

19. Let $[C_n]_{n<\omega}$ be any sequence of groups, let $P = \prod_{n<\omega} C_n$ and $S = \sum_{n<\omega} C_n$. Denote by $\overline{S}$ the closure of S in the n-adic topology on P.

(a) Prove that $x = \langle x_n \rangle \in \overline{S}$ if and only if $x_n = k_n! y_n$ where $k_n \to \infty$ as $n \to \infty$.

(b) Prove that $\overline{S}$ is a pure subgroup of P.

20. Show that $\prod_{\aleph_0} Z/\sum_{\aleph_0} Z \simeq \prod_{\aleph_0} E(Z) + \sum_{2^{\aleph_0}} Q$. (<u>Hint</u>: Let $\prod_{\aleph_0} Z/\sum_{\aleph_0} Z = A + D$ where A is reduced and D is divisible. From the exactness of $\mathrm{Ext}(Q/Z, \sum_{\aleph_0} Z) \rightarrowtail \prod_{\aleph_0} \mathrm{Ext}(Q/Z, Z) \twoheadrightarrow \mathrm{Ext}(Q/Z, A + D)$, deduce that $\prod_{\aleph_0} E(Z) \simeq E(S) + A$ and that $\mathrm{rank}(A/pA) = 2^{\aleph_0}$ for each prime p. Then show that A contains a pure free dense subgroup F of cardinality $2^{\aleph_0}$ and that $A = E(F) = \prod_{\aleph_0} E(Z)$.)

21. Let E = E(T) where T is a reduced torsion group with $\bigcap_{n<\omega} nT \neq 0$. Prove that E is not algebraically compact.

V. DIRECT SUMS OF COUNTABLE p-GROUPS AND HILL'S VERSION OF ULM'S THEOREM

In this chapter all groups will be p-primary for a fixed but arbitrary prime p (unless otherwise stated).

Definition. If S is a p-group such that $pS = 0$, then we use the notation dimS to indicate the vector space dimension of S over the field $Z/(p)$. Note also that dimS = rankS.

It is appropriate at this time to refine our notions of p^nG and the height function $h^G(x)$.

Definition. For each ordinal number α, we define $p^{\alpha+1}G = p(p^{\alpha}G)$ and $p^{\alpha}G = \bigcap_{\beta<\alpha} p^{\beta}G$ if α is a limit ordinal. Since a group is a set, there is some ordinal number λ after which all $p^{\alpha}G$ are equal and, hence, equal to the maximal divisible subgroup of G. If G is reduced, we call the first λ such that $p^{\lambda}G = 0$ the length of G. Furthermore, we define for $x \in G$,

$$h^G(x) = \begin{cases} \alpha, & \text{if } x \in p^{\alpha}G \text{ but } x \notin p^{\alpha+1}G \\ \infty, & \text{if } x \text{ is in the maximal divisible subgroup of } G \end{cases}$$

We write $p^{\alpha}G[p]$ to mean $(p^{\alpha}G)[p]$ and note that these subgroups of G are obviously invariant subgroups of G. Then, for $p^{\alpha}G[p] = p^{\alpha+1}G[p] + S_{\alpha}$, it follows that $f_G(\alpha) = \dim S_{\alpha}$ is an invariant of G. Following Kaplansky [69], we call $f_G(\alpha)$ the α-th Ulm invariant of G.

If G is a direct sum $G = \sum_{i\in I}\{g_i\}$ of cyclic p-groups, then $f_G(n)$, $0 \leq n < \omega$, is just the cardinal number of $\{g_i\}$ of order p^{n+1} and $f_G(\alpha) = 0$ for $\alpha \geq \omega$. Thus, for the class of p-groups which are direct sums of cyclic groups, the cardinal numbers $f_G(\alpha)$ form a complete set of cardinal invariants. To show that (unfortunately) the above invariants are not sensitive enough in general to distinguish between p-groups, we refer the reader to

Exercise 14 (Chapter III) to the group $B = \sum_{n<\omega} Z(p^n)$ and its torsion completion $\overline{B}$. It is left as an exercise to show that $f_B(n) = f_{\overline{B}}(n) = 1$ for $0 \leq n < \omega$ and $f_B(\alpha) = f_{\overline{B}}(\alpha) = 0$ for $\alpha \geq \omega$. However, $B \neq \overline{B}$. To find groups of the same cardinality having the same Ulm invariants but not isomorphic, let $G = B + C$ and $H = \overline{B} + C$ where $C = \sum_{n<\omega} \sum_{2^{\aleph_0}} Z(p^n)$. Then G and H are the desired groups.

Following Hill [54] we introduce the notion of relative Ulm invariants for considering a group and one of its subgroups.

Definition. Let G be a reduced p-group, let $p^\alpha G[p] = p^{\alpha+1}G[p] + S_\alpha$ for each α and let A be a subgroup of G. Define $S_\alpha(A)$ by $S_\alpha(A) = \{x \in S_\alpha : x + a \in p^{\alpha+1}G \text{ for some } a \in A\}$ and further define the α-th Ulm invariant of G relative to A by

$$F_\alpha(G,A) = \dim(S_\alpha/S_\alpha(A)).$$

Note that when $A = 0$, then $f_G(\alpha) = F_\alpha(G,0)$ for all α.

In order to quiet suspicions that $F_\alpha(G,A)$ might depend upon the choice of S_α, we state emphatically in our first result on $F_\alpha(G,A)$ that it does not and leave its proof as an exercise.

Proposition 54. Let G, A and $F_\alpha(G,A)$ be as above. Then $F_\alpha(G,A)$ is independent of the choice of the direct summand S_α in the above decomposition $p^\alpha G[p] = p^{\alpha+1}G[p] + S_\alpha$.

Definition. Let G be a p-group and let A be a subgroup of G. Then $x \in G$ is called proper with respect to A if x has maximal height among the elements of the coset $x + A$.

Proposition 55. Suppose that the invariants of G relative to A are the same as the invariants of $\overline{G}$ relative to $\overline{A}$. Then there exists an element in G[p] that has height exactly α and is proper with respect to A if and only if there exists an element in $\overline{G}[p]$ that has height exactly α and is proper with respect to $\overline{A}$.

Proof. Suppose that $x \in G[p]$ has height α and is proper with respect to A. Recall that x is proper with respect to A if x has maximal height among the elements of the coset $x + A$. Let $x = y + z$ where $y \in S_\alpha$ and $z \in p^{\alpha+1}G[p]$. Then y is proper with respect to A and, therefore, is not contained in $S_\alpha(A)$. Thus, $F_\alpha(G,A) \neq 0$. Conversely, if $F_\alpha(G,A) \neq 0$, there exists an element $x \in G[p]$ that has height α and is proper with respect to A. Since $F_\alpha(G,A) = F_\alpha(\overline{G},\overline{A})$, the proposition follows.

The next result shows that equality of the relative invariants $F_\alpha(G,A) = F_\alpha(\overline{G},\overline{A})$ is not destroyed by finite extensions of the subgroups A and $\overline{A}$, respectively, provided the new subgroups correspond under a height preserving isomorphism. As usual, all heights are computed in the containing groups G and $\overline{G}$ (not with respect to subgroups).

Proposition 56. Let $F_\alpha(G,A) = F_\alpha(\overline{G},\overline{A})$ and suppose that π is a height-preserving isomorphism from A onto $\overline{A}$. If $\phi: B \twoheadrightarrow \overline{B}$ is a finite extension of π (that is, B/A is finite) and if ϕ is a height-preserving isomorphism from B onto $\overline{B}$ where $A \subseteq B \subseteq G$ and $\overline{A} \subseteq \overline{B} \subseteq \overline{G}$, then $F_\alpha(G,B) = F_\alpha(\overline{G},\overline{B})$.

Proof. Let $p^\alpha G[p] = p^{\alpha+1}\overline{G}[p] + S_\alpha$ and $p^\alpha G[p] = p^{\alpha+1}\overline{G}[p] + T_\alpha$. Since $F_\alpha(G,A) = F_\alpha(\overline{G},\overline{A})$, there is an isomorphism from $S_\alpha/S_\alpha(A)$ onto $T_\alpha/T_\alpha(\overline{A})$. Define $B^*_\alpha = \{x \in p^\alpha G \cap B: px \in p^{\alpha+1}G\}$. The correspondence $x \to b$ where $x + b \in p^{\alpha+1}G$ induces an isomorphism θ from $S_\alpha(B)$ onto $B^*_\alpha/(p^{\alpha+1}G \cap B)$. In the same way, we have a natural isomorphism $\overline{\theta}$ from $T_\alpha(\overline{B})$ onto $\overline{B}^*_\alpha/(p^{\alpha+1}\overline{G} \cap \overline{B})$. Let ϕ^* be the isomorphism $B^*_\alpha/(p^{\alpha+1}G \cap B) \rightarrowtail\!\!\!\twoheadrightarrow \overline{B}^*_\alpha/(p^{\alpha+1}\overline{G} \cap \overline{B})$ induced by the height-preserving isomorphism $\phi: B \rightarrowtail\!\!\!\twoheadrightarrow \overline{B}$. Now observe that the isomorphism $f = \overline{\theta}^{-1}\phi^*\theta: S_\alpha(B) \rightarrowtail\!\!\!\twoheadrightarrow T_\alpha(\overline{B})$ maps $S_\alpha(A)$ onto $T_\alpha(\overline{A})$. Thus, we further obtain an isomorphism from $S_\alpha(B)/S_\alpha(A)$ onto $T_\alpha(\overline{B})/T_\alpha(\overline{A})$. Since there is a natural injection of $S_\alpha(B)/S_\alpha(A)$ into $\{B,p^{\alpha+1}G\}/\{A,p^{\alpha+1}G\}$, it follows that $S_\alpha(B)/S_\alpha(A)$ is finite. Similarly, $T_\alpha(\overline{B})/T_\alpha(\overline{A})$ is finite. Hence, $S_\alpha/S_\alpha(B) \simeq T_\alpha/T_\alpha(\overline{B})$ and the proof is finished.

Definition. Again following Hill [54], we say that a subgroup A of a group G is nice in G if $p^\alpha(G/A) = \{p^\alpha G,A\}/A$ for all α. Observe that A is nice in G if and only if, for $x \in G$, $h^{G/A}(x + A) = h^G(x + a)$ for some $a \in A$.

If one looks closely at the Kaplansky-Macky proof of Ulm's Theorem for countable p-groups, one sees that Hill [54] is able to replace the finite subgroups in the Kaplansky-Macky proof [see pp. 26-30, 69] by the notion of nice subgroup.

Proposition 57. The subgroup A of G is nice if and only if each coset $x + A$ contains an element $x + a$ that is proper with respect to A.

Proof. If A is nice, there exists $a \in A$ such that $h^{G/A}(x + A) = h^G(x + a)$. Obviously, $x + a$ is proper with respect to A. Conversely, suppose that each coset $x + A$ contains an element that is proper with respect to A. Then we can verify that $p^\alpha(G/A) = \{p^\alpha G, A\}/A$ by induction on α; it is enough to prove that $p^\alpha(G/A) \subseteq \{p^\alpha G, A\}$. Assume that $p^\alpha(G/A) \subseteq \{p^\alpha G, A\}/A$ for all $\alpha < \beta$. Let $x + A \in p^\beta(G/A)$. If $\beta - 1$ exists, then $x + A = p(y + A)$ where $y + A \in p^{\beta-1}(G/A)$. By the induction hypothesis, $y \in \{p^{\beta-1}G, A\}$. Therefore, $x \in \{p^\beta G, A\}$, since $py \in \{p^\beta G, A\}$ and since $x = py + a$, $a \in A$. If β is a limit ordinal, choose x proper with respect to A. If $\alpha < \beta$, there exists by the induction hypothesis $a_\alpha \in A$ such that $h^G(x + a_\alpha) \geq \alpha$. Thus, $h^G(x) \geq \alpha$ for each $\alpha < \beta$ and $x \in p^\beta G$.

Immediate consequences of Proposition 57 are that finite subgroups, direct summands, $p^\alpha G$ (for any α) and subgroups H of G with $p^\omega(G/H) = 0$ are nice subgroups of the p-group G. That nice subgroups encompass even a wider class of subgroups is shown by the next two elementary propositions which we leave as exercises.

Proposition 58. If A is a nice subgroup of G and if $A \subseteq B \subseteq G$ such that B/A is nice in G/A, then B is nice in G.

Proposition 59. If A_i is nice in G_i, then $A = \sum A_i$ is nice in $G = \sum G_i$.

Proposition 60. Let β denote an arbitrary ordinal and let A be a subgroup of the group G. Then A is a nice subgroup G if and only if $\{A, p^\beta G\}/p^\beta G$ is a nice subgroup of $G/p^\beta G$ and $A \cap p^\beta G$ is a nice subgroup of $p^\beta G$.

Proof. We first prove the sufficiency. If $\alpha \leq \beta$, then $p^\alpha(G/p^\beta G/\{A, p^\beta G\}/p^\beta G) = \{p^\alpha G/p^\beta G, \{A, p^\beta G\}/p^\beta G\}/(\{A, p^\beta G\}/p^\beta G) =$

$(\{A,p^{\alpha}G\}/p^{\beta}G)/(\{A,p^{\beta}G\}/p^{\beta}G)$, since $\{A,p^{\beta}G\}/p^{\beta}G$ is nice in $G/p^{\beta}G$. Therefore, if $\alpha \leq \beta$, $p^{\alpha}(G/A/(\{A,p^{\alpha}G\}/A)) \simeq p^{\alpha}(G/\{A,p^{\alpha}G\}) \simeq p^{\alpha}((G/p^{\beta}G)/(\{A,p^{\alpha}G\}/p^{\beta}G) \simeq p^{\alpha}((G/p^{\beta}G/\{A,p^{\beta}G\}/p^{\beta}G)/(\{A,p^{\beta}G\}/p^{\alpha}G/\{A,p^{\beta}G\}/p^{\beta}G) = p^{\alpha}(X/p^{\alpha}X) = 0$, where $X = G/p^{\beta}G/\{A,p^{\beta}G\}/p^{\beta}G$. Hence, $p^{\alpha}(G/A) = \{A,p^{\alpha}G\}/A$ if $\alpha \leq \beta$.

Now suppose that $\alpha = \beta + \gamma$. We have that $p^{\gamma}((p^{\beta}G/A \cap p^{\beta}G)/(\{p^{\alpha}G,A \cap p^{\beta}G\}/A \cap p^{\beta}G)) = 0$ since $p^{\gamma}(p^{\beta}G/A \cap p^{\beta}G) = \{p^{\alpha}G,A \cap p^{\beta}G\}/A \cap p^{\beta}G$. Thus, $p^{\gamma}((\{A,p^{\beta}G\}/A)/(\{A,p^{\alpha}G\}/A)) = 0$ and $p^{\alpha}(G/A) = p^{\gamma}(p^{\beta}(G/A)) = p^{\gamma}(\{A,p^{\beta}G\}/A) = \{A,p^{\alpha}G\}/A$.

The proof of the necessity is very similar to the above and thus its proof is left to the reader.

In 1935, Zippen [138] improved upon Ulm's proof [130] that the Ulm invariants are adequate to classify the class of countable reduced p-groups. His proof actually revealed more in that Zippen showed that if G and $\overline{G}$ are countable and have the same Ulm invariants, then any isomorphism from $p^{\alpha}G$ onto $p^{\alpha}\overline{G}$ can be extended to an isomorphism from G onto $\overline{G}$. Hill [54] has extended Zippen's Theorem as follows:

<u>Theorem 61</u>. (Hill) Let A and $\overline{A}$ be nice subgroups of G and $\overline{G}$, respectively. Suppose that G and $\overline{G}$ have the same invariants relative to A and $\overline{A}$, respectively. If G/A and $\overline{G}/\overline{A}$ are countable, then any height-preserving isomorphism from A onto $\overline{A}$ can be extended to an isomorphism from G onto $\overline{G}$.

<u>Proof</u>. Let π_o be a height-preserving isomorphism from A onto $\overline{A}$. Let $G = \{A,x_1,x_2,\ldots\}$ and $\overline{G} = \{\overline{A},\overline{x}_1,\overline{x}_2,\ldots\}$. Suppose that we have an ascending finite sequence

$$\pi_o \leq \pi_1 \leq \cdots \leq \pi_n$$

of height-preserving isomorphisms such that each goes from a subgroup of G onto a subgroup of $\overline{G}$ and each is a finite extension of π_o. Assume that x_i is in the domain of π_{2i-1} if $2i-1 \leq n$ and x_i is in the image of π_{2i} if $2i \leq n$. We wish to extend π_n to a height-preserving isomorphism π_{n+1} such that the above conditions are still valid if $n+1$ replaces n. Because of the symmetry between G and $\overline{G}$ and the reversibility of an isomorphism, it is enough to consider the case that n is even; set $n = 2k$. Let $\pi_n : B \rightarrowtail\!\!\!\twoheadrightarrow \overline{B}$. We intend to extend π_n to a height-preserving isomorphism from $\{B, x_{k+1}\}$ into $\overline{G}$. If $x_{k+1} \in B$, we can take $\pi_{n+1} = \pi_n$ so we assume $x_{k+1} \notin B$. There is no loss in generality in assuming that $px_{k+1} \in B$. Since any finite subgroup is nice, B/A is nice in G/A. Thus, B is a nice subgroup of G by Proposition 58. Likewise, $\overline{B}$ is a nice subgroup of $\overline{G}$. This means that we may assume that x_{k+1} is proper with respect to B. By Proposition 56, $F_\alpha(G,B) = F_\alpha(\overline{G},\overline{B})$ for each α. The rest of the proof follows that of Theorem 14 [pp. 29-30, 69]. For simplicity of notation, set $x = x_{k+1}$ and $h^G(x) = \alpha$.

<u>Case 1</u>. $h^G(px) > \alpha + 1$. Let $px = py$ where $y \in p^{\alpha+1}G$. Then $z = x - y$ is in G[p], has height exactly α and is proper with respect to B. By Proposition 55, there exists $\overline{z} \in \overline{G}[p]$ that has height exactly α and is proper with respect to $\overline{B}$. Choose $\overline{y} \in p^{\alpha+1}\overline{G}$ such that $p\overline{y} = \pi_n(px)$; this is possible since π_n is height-preserving. Now π_n can be extended in the desired manner by mapping x onto $\overline{y} + \overline{z}$.

<u>Case 2</u>. $h^G(px) = \alpha + 1$. Choose $\overline{y} \in p^{\alpha}\overline{G}$ such that $p\overline{y} = \pi_n(px)$. If $\overline{y} \in \overline{B}$, then $\pi_n(b) = \overline{y}$ for some $b \in B \cap p^\alpha G$. Since $b \in p^\alpha G$, $x - b$ is also proper with respect to B. However, $px = pb$ so $x - b$ satisfies Case 1, $h^G(p(x - b)) > \alpha + 1$. Thus, we may assume that $\overline{y} \notin \overline{B}$. A similar argument shows that the assumption that $\overline{y}$ is not proper with respect to $\overline{B}$ returns us to Case 1. Hence, we may assume that $\overline{y}$ is proper with respect to $\overline{B}$. Mapping x onto $\overline{y}$ produces a height-preserving isomorphism from {B,x} into G.

We have shown the existence of a sequence

$$\pi_0 \leq \pi_1 \leq \ldots \leq \pi_n \leq \ldots$$

of isomorphisms from subgroups of G to subgroups of $\overline{G}$ such that x_i is in the domain of π_{2i-1} and x_i is in the image of π_{2i}. Clearly, $\pi = \sup_i\{\pi_i\}$ is an isomorphism from G onto $\overline{G}$, that is, $\pi(x) = \pi_i(x)$ where x is in the domain of π_i.

We now present Hill's axiomatic evolution of "Ulm's Theorem" for reduced p-groups.

<u>Axiom 1</u>. G is countable.

<u>Axiom 2</u>. G is a direct sum of countable groups; hereafter called d.s.c. groups.

<u>Axiom 3</u>. G has a collection C of nice subgroups satisfying the following conditions:

(0) $0 \in C$.

(1) C is closed with respect to ascending unions.

(2) If $A \in C$ and if H is a subgroup of G such that $\{A,H\}/A$ is countable, there exists $B \in C$ such that $\{A,H\} \subseteq B$ and B/A is countable.

In 1933, Ulm [130] proved that the "Ulm Invariants" suffice to classify Axiom 1 groups and (as mentioned above) Zippen [138] improved the situation in 1935. The most elegant account is due to Kaplansky and Macky [69]. Not until 1958 was the situation further improved. At this time Kolettis [75] showed that the theorem could be extended to reduced d.s.c. p-groups (Axiom 2 groups). A much simpler account of this result was given by Hill [49] in 1966. At this point the class of groups (reduced d.s.c. p-groups), for which the Ulm invariants served to classify them, is not restricted by cardinality but must have length not exceeding Ω (= first uncountable ordinal). Early in 1967, Walker and Parker [109] were able to extend the theorem to Axiom 3 groups whose length was less than $\Omega\omega$. Finally, in the summer of 1967, Hill [54] set forth the concepts of relative invariants, nice subgroups and third axiom of countability, and succeeded in proving that the Ulm invariants suffice to classify Axiom 3 groups and, moreover, in some sense (see Theorem 70) this class of groups is maximal. We extract the main part of the induction step in Hill's proof and make it a preliminary technical lemma.

<u>Lemma 62</u>. Suppose that A and $\overline{A}$ are nice subgroups of G and $\overline{G}$, respectively, such that $F_\alpha(G,A) = F_\alpha(\overline{G},\overline{A})$ for all α and suppose that G/A and $\overline{G}/\overline{A}$

satisfy the third axiom of countability. Let $\mathcal{C}$ and $\overline{\mathcal{C}}$ be collections of nice subgroups of G/A and $\overline{G}/\overline{A}$, respectively, that satisfy the conditions of Axiom 3. Let $p^{\alpha}G[p] = p^{\alpha+1}G[p] + S_{\alpha}$ and $p^{\alpha}\overline{G}[p] = p^{\alpha+1}\overline{G}[p] + T_{\alpha}$ and further let $S_{\alpha} = S_{\alpha}(A) + P_{\alpha}$ and $T_{\alpha} = T_{\alpha}(\overline{A}) + Q_{\alpha}$ for each α. Furthermore, let ψ_{α} be an isomorphism from P_{α} onto Q_{α} (recall that $F_{\alpha}(G,A) = F_{\alpha}(\overline{G},\overline{A})$). Suppose that $A \subseteq B \subseteq G$ and $\overline{A} \subseteq \overline{B} \subseteq \overline{G}$ with $B/A \in \mathcal{C}$ and $\overline{B}/\overline{A} \in \overline{\mathcal{C}}$ such that ψ_{α} maps $P_{\alpha}(B) = S_{\alpha}(B) \cap P_{\alpha}$ onto $Q_{\alpha}(\overline{B}) = T_{\alpha}(\overline{B}) \cap Q_{\alpha}$. If π is a height-preserving isomorphism of B onto $\overline{B}$ which takes A onto $\overline{A}$ and if $x \in G$ and $\overline{x} \in \overline{G}$, there are subgroups $C \supseteq B$ and $\overline{C} \supseteq \overline{B}$ of G and $\overline{G}$, respectively, and a height-preserving isomorphism $\pi_1 : C \rightarrowtail\!\!\!\twoheadrightarrow \overline{C}$ which extends π such that

(a) $C/A \in \mathcal{C}$ and $\overline{C}/\overline{A} \in \overline{\mathcal{C}}$.

(b) For each α, ψ_{α} maps $P_{\alpha}(C) = S_{\alpha}(C) \cap P_{\alpha}$ onto $Q_{\alpha}(\overline{C}) = T_{\alpha}(C) \cap Q_{\alpha}$.

(c) $x \in C$ and $\overline{x} \in \overline{C}$.

<u>Proof</u>. Recall that $S_{\alpha}(A) = \{x \in S_{\alpha} : x + a \in p^{\alpha+1}G \text{ for some } a \in A\}$; $S_{\alpha}(B)$, $T_{\alpha}(\overline{A})$ and $T_{\alpha}(\overline{B})$ are defined analogously. We have that $S_{\alpha}(B) = S_{\alpha}(A) + P_{\alpha}(B)$ and $T_{\alpha}(\overline{B}) = T_{\alpha}(\overline{A}) + Q_{\alpha}(\overline{B})$ for all α. Moreover, the fact that ψ_{α} maps $P_{\alpha}(B)$ onto $Q_{\alpha}(\overline{B})$ implies that

$$S_{\alpha}/S_{\alpha}(B) \simeq P_{\alpha}/P_{\alpha}(B) \simeq Q_{\alpha}/Q_{\alpha}(\overline{B}) \simeq T_{\alpha}/T_{\alpha}(\overline{B})$$

and $F_{\alpha}(G,B) = F_{\alpha}(\overline{G},\overline{B})$ for all α. We shall construct subgroups $C \supseteq B$ and $\overline{C} \supseteq \overline{B}$ and a height-preserving $\pi_1 : C \rightarrowtail\!\!\!\twoheadrightarrow \overline{C}$ such that conditions (a)-(c) are satisfied. First we take care of condition (c). It is clear from the proof of Theorem 61 that there exists a height-preserving isomorphism $\phi_0 : B_0 \rightarrowtail\!\!\!\twoheadrightarrow \overline{B}_0$ such that ϕ_0 is a finite extension of π and such that $x \in B_0$, $x \in \overline{B}$. We now make an important observation: if $A \subseteq E \subseteq E' \subseteq G$, then $\sum_{\alpha} |P_{\alpha}(E')/P_{\alpha}(E) - 0| \leq |E'/E|$ where α ranges over the ordinals. Let M_{α} be a set of representatives for the nonzero cosets of $P_{\alpha}(E')$ with respect to $P_{\alpha}(E)$; if $P_{\alpha}(E) = P_{\alpha}(E')$, then M_{α} is void. For each $x \in M_{\alpha}$, choose one and only one element $e'(x)$ in E' such that $x + e'(x) \in p^{\alpha+1}G$. Define a function $f_{\alpha} : M_{\alpha} \to E'/E$ by $f_{\alpha}(x) = e'(x) + E$. Then f_{α} is an injection and $f_{\alpha}(M) \cap f_{\beta}(M_{\beta}) = \phi$ if $\alpha \neq \beta$. This reveals what we wanted to observe.

Since B_o/B and $\overline{B}_o/\overline{B}$ are countable, there exist subgroups C_1 and $\overline{C}_1$ of G and $\overline{G}$, respectively, such that:

(i) $B_o \subseteq C_1$ and $\overline{B}_o \subseteq \overline{C}_1$.

(ii) $\psi_\alpha(P_\alpha(B_o)) \subseteq Q_\alpha(\overline{C}_1)$ and $\psi_\alpha(P_\alpha(C_1)) \supseteq Q_\alpha(\overline{B}_o)$, for each α.

(iii) C_1/B and $\overline{C}_1/\overline{B}$ are countable and $C_1/A \in \mathcal{C}$ and $\overline{C}_1/\overline{B} \in \overline{\mathcal{C}}$.

We remark that the above observation is needed in order that we might have (ii) and (iii) hold simultaneously. Let $C_1 = \{B, x_1^{(1)}, x_2^{(1)}, \ldots, x_n^{(1)}, \ldots\}$ and $\overline{C}_1 = \{B, \overline{x}_1^{(1)}, \overline{x}_2^{(1)}, \ldots, \overline{x}_n^{(1)}, \ldots\}$. There exists a height-preserving isomorphism $\phi_1 \colon B_1 \rightarrowtail\!\!\!\!\twoheadrightarrow \overline{B}_1$ such that ϕ_1 is a finite extension of ϕ_o and such that $x_1^{(1)} \in B_1$ and $\overline{x}_1^{(1)} \in B_1$. There exist subgroups $C_2 \supseteq C_1$ and $\overline{C}_2 \supseteq \overline{C}_1$ of G and $\overline{G}$, respectively, such that conditions (i)-(iii) are satisfied when B_o, $\overline{B}_o$, C_1 and $\overline{C}_1$ are replaced by B_1, $\overline{B}_1$, C_2 and $\overline{C}_2$, respectively. Let $C_2 = \{B, x_1^{(2)}, x_2^{(2)}, \ldots, x_n^{(2)}, \ldots\}$ and $\overline{C}_2 = \{B, \overline{x}_1^{(2)}, \overline{x}_2^{(2)}, \ldots, \overline{x}_n^{(2)}, \ldots\}$. There exists a height-preserving isomorphism $\phi_2 \colon B_2 \rightarrowtail\!\!\!\!\twoheadrightarrow \overline{B}_2$ such that ϕ_2 is a finite extension of ϕ_1 and such that $x_j^{(i)} \in B$ and $\overline{x}_j^{(i)} \in \overline{B}$ if $i \leq 2$ and $j \leq 2$. There exists subgroups $C_3 \supseteq C_2$ and $\overline{C}_3 \supseteq \overline{C}_2$ of G and $\overline{G}$, respectively, such that conditions (i)-(iii) are satisfied when B_o, $\overline{B}_o$, C_1 and $\overline{C}_1$ are replaced by B_2, $\overline{B}_2$, C_3 and $\overline{C}_3$, respectively. Continue in this manner. Set $C = \bigcup_{n<\omega} B_n = \bigcup_{n<\omega} C_n$ and $\overline{C} = \bigcup_{n<\omega} \overline{B}_n = \bigcup_{n<\omega} \overline{C}_n$. Define $\pi_1 \colon C \rightarrowtail\!\!\!\!\twoheadrightarrow \overline{C}$ by $\pi_1 = \sup_{n<\omega}\{\phi_n\}$. Then π_1 is a height-preserving isomorphism that extends π. We claim that conditions (a)-(c) are satisfied; we took care of condition (c) in the initial step of the extension of π. Since $C_i/A \in \mathcal{C}$ and $\overline{C}_i/\overline{A} \in \overline{\mathcal{C}}$, condition (a) follows from postulate (1) of Axiom 3. Condition (b) follows from $\psi_\alpha(P_\alpha(B_i)) \subseteq Q_\alpha(\overline{C}_{i+1})$ and $\psi_\alpha(P_\alpha(C_{i+1})) \supseteq Q_\alpha(\overline{B}_i)$.

Hill's version of Ulm's Theorem is contained in the following result.

<u>Theorem 63</u>. (Hill) Suppose that A and $\overline{A}$ are nice subgroups of G and $\overline{G}$, respectively, such that $F_\alpha(G,A) = F_\alpha(\overline{G},\overline{A})$ for all α. If G/A and $\overline{G}/\overline{A}$ satisfy the third axiom of countability, then any height-preserving isomorphism from A onto $\overline{A}$ can be extended to an isomorphism from G onto $\overline{G}$.

<u>Proof</u>. Let $G = \{A, x_\lambda\}_{\lambda\in\Lambda}$ and $\overline{G} = \{\overline{A}, \overline{x}_\lambda\}_{\lambda\in\Lambda}$; there is no loss of

generality in using a common index set since the x_λ's need not be distinct. If Λ is countable, Theorem 61 applies and the proof is finished. Thus, assume that Λ is uncountable; for convenience of notation, let Λ be the set of ordinal numbers less than Γ where Γ is the first ordinal having the appropriate cardinality. Let $\mathcal{C}$ and $\overline{\mathcal{C}}$ be collections of nice subgroups of G/A and $\overline{G}/\overline{A}$, respectively, that satisfy the conditions of Axiom 3. As usual, we let $p^\alpha G[p] = p^{\alpha+1}G[p] + S_\alpha$ and $p^\alpha \overline{G}[p] = p^{\alpha+1}\overline{G}[p] + T_\alpha$ for each ordinal α. Let $S_\alpha = S_\alpha(A) + P_\alpha$ and $T_\alpha = T_\alpha(\overline{A}) + Q_\alpha$. Since $F_\alpha(G,A) = F_\alpha(\overline{G},\overline{A})$, there is an isomorphism ψ_α from P_α onto Q_α. Finally, if $A \subseteq B \subseteq G$ and $\overline{A} \subseteq \overline{B} \subseteq \overline{G}$, define $P_\alpha(B) = S_\alpha(B) \cap P_\alpha$ and $Q_\alpha(\overline{B}) = T_\alpha(\overline{B}) \cap Q_\alpha$. Then $S_\alpha(B) = S_\alpha(A) + P_\alpha(B)$ and $T_\alpha(\overline{B}) = T_\alpha(\overline{A}) + Q_\alpha(\overline{B})$.

Suppose that π_0 is a height-preserving isomorphism from A onto $\overline{A}$. Assume that $\mu < \Gamma$ and that we have an ascending chain $\pi_0 \leq \pi_1 \leq \ldots \leq \pi_\lambda \leq \ldots$, $\lambda < \mu$, of height-preserving isomorphisms $\pi_\lambda : A_\lambda \rightarrowtail\!\!\!\twoheadrightarrow \overline{A}_\lambda$ such that, for $\lambda < \mu$,

(a) $A_\lambda/A \in \mathcal{C}$ and $\overline{A}_\lambda/\overline{A} \in \overline{\mathcal{C}}$.

(b) For each α, ψ_α maps $P_\alpha(A_\lambda)$ onto $Q_\alpha(\overline{A}_\lambda)$.

(c) $x_{\lambda-1} \in A_\lambda$ and $\overline{x}_{\lambda-1} \in \overline{A}_{\lambda-1}$ if $\lambda-1$ exists.

The conditions are trivially met when $\mu = 1$, $\lambda = 0$; note that $A_0 = A$ and $\overline{A}_0 = \overline{A}$. We wish to construct a height-preserving isomorphism $\pi_\mu : A_\mu \rightarrowtail\!\!\!\twoheadrightarrow \overline{A}_\mu$ that extends π_λ, for $\lambda < \mu$, in such a way that conditions (a)-(c) are easily verified for $\lambda \leq \mu$. If μ is a limit ordinal, define $A_\mu = \bigcup_{\lambda<\mu} A_\lambda$, $\overline{A}_\mu = \bigcup_{\lambda<\mu} \overline{A}_\lambda$ and let $\pi_\mu : A_\mu \rightarrowtail\!\!\!\twoheadrightarrow \overline{A}_\mu$ be defined by $\pi_\mu = \sup_{\lambda<\mu}\{\pi_\lambda\}$. Then π_μ is height-preserving and conditions (a)-(c) are easily verified for $\lambda \leq \mu$. If $u-1$ exists we simply apply Lemma 62 to obtain π_μ, A_μ and $\overline{A}_\mu$ satisfying (a)-(c).

We have now the existence of an ascending chain $\pi_0 \leq \pi_1 \leq \ldots \leq \pi_\lambda \leq \ldots$, $\lambda < \Gamma$, of isomorphisms $\pi_\lambda : A_\lambda \rightarrowtail\!\!\!\twoheadrightarrow \overline{A}_\lambda$ where $A_\lambda \subseteq G$, $\overline{A}_\lambda \subseteq \overline{G}$, $G = \cup A_\lambda$ and $\overline{G} = \cup \overline{A}_\lambda$. Clearly, $\pi = \sup_{\lambda<\Gamma}\{\pi_\lambda\}$ is an isomorphism from G onto $\overline{G}$ that extends π_0.

We next consider some interesting properties of Axiom 2 and Axiom 3 groups after which we establish the existence of Axiom 3 groups of arbitrary length and the "maximality" of this latest version of Ulm's Theorem.

Theorem 64. If G_i is an Axiom 3 group for $i \in I$, then $G = \sum_{i \in I} G_i$ is also an Axiom 3 group.

Proof. Let $\mathcal{C}_i$ be a collection of nice subgroups of G_i that satisfies the conditions of Axiom 3. Let $\mathcal{C}$ be the collection of subgroups N of G such that $N = \sum_{i \in I} N_i$ where $N_i \in \mathcal{C}_i$. Proposition 59 states that such an N is a nice subgroup of G. Obviously, $0 \in \mathcal{C}$ and $\mathcal{C}$ is closed with respect to ascending unions. Suppose that $N = \sum N_i \in \mathcal{C}$ and that H is a subgroup of G such that $|\{H,N\}/N| \leq \aleph_0$. Let H_i denote the projection of $G = \sum G_i$ onto G_i. Then $H_i \subseteq N_i$ for all but a countable number of i because $\{H,N\}/N$ is countable. Since $|\{H_i,N_i\}/N_i| \leq \aleph_0$, there exists $N_i' \in \mathcal{C}_i$ such that $\{H_i,N_i\} \subseteq N_i'$ and N_i'/N_i is countable. It follows that there exists $N' = \sum N_i' \in \mathcal{C}$ such that $N' \supseteq \{H,N\}$ and N'/N is countable.

The infinite juggling technique used in the proof of Theorem 63 was initiated by Kaplansky [71] in proving the following remarkable result. In this theorem (only) we allow the groups to be arbitrary.

Theorem 65. (Kaplansky) A direct summand of a d.s.c. group is again a d.s.c.

Proof. Suppose that $C = \sum_{i \in I} C_i = A + B$ where each C_i is countable and where I is an initial segment of the ordinal numbers beginning with 0. We wish to show that A is a d.s.c.. So let π be the natural projection of C onto A. We first argue that I can be written as an ascending union $I = \bigcup_{\alpha \in I} I_\alpha$ satisfying:

(i) $\alpha \in I_{\alpha+1}$.

(ii) $I_\alpha = \bigcup_{\beta<\alpha} I_\beta$ if α is a limit ordinal.

(iii) $|I_{\alpha+1} - I_\alpha| \leq \aleph_0$.

(iv) If ρ_α is the natural projection of C onto $\sum_{I_\alpha} C_i$, then $\pi\rho_\alpha$ is a projection of A onto $A_\alpha = A \cap \sum_{I_\alpha} C_i$.

Let $I_o = \phi$ and $A_o = 0$ and suppose $[I_\alpha]_{\alpha<\gamma}$ have been constructed so that (i)-(iv) hold. If γ is a limit, set $I_\gamma = \bigcup_{\alpha<\gamma} I_\alpha$. Then clearly $[I_\alpha]_{\alpha\leq\gamma}$ satisfies (i)-(iv). Now suppose that $\gamma = \beta + 1$. For each positive integer n, we obtain a subset $L_n \supseteq I_\beta$ such that $L_1 = I_\beta \cup [\beta]$, $L_n \subseteq L_{n+1}$ and $A \cap \sum_{L_1} C_i \subseteq \pi(\sum_{L_1} C_i) \subseteq \ldots \subseteq \pi(\sum_{L_n} C_i) \subseteq A \cap \sum_{L_{n+1}} C_i \subseteq \ldots$. If $|L_n - I_\beta| \leq \aleph_0$, then we may choose L_{n+1} so that $|L_{n+1} - I_\beta| \leq \aleph_0$ since $\pi(\sum_{L_n} C_i)$ is countably generated modulo $A \cap \sum_{I_\beta} C_i$. Set $I_{\beta+1} = \bigcup_{n<\omega} L_n$ and observe that (i)-(iii) are still satisfied. If $a \in A$, then for some $n < \omega$, $\rho_{\beta+1}(a) = x \in \sum_{L_n} C_i$ and $\pi(x) \in \sum_{L_{n+1}} C_i$ which shows that $\rho_{\beta+1}\pi(x) = \pi(x)$. Hence, $\pi\rho_{\beta+1} = \pi(x)$ and $\pi\rho_{\beta+1}\pi\rho_{\beta+1}(a) = \pi(\rho_{\beta+1}\pi(x)) = \pi(\pi(x)) = \pi(x)$ proving that (iv) also holds for the collection $[I_\alpha]_{\alpha\leq\gamma}$. Thus, I can be written as an ascending union $I = [I_\alpha]_{\alpha\in I}$ satisfying (i)-(iv) above.

Now (iii) and (iv) show that $A_{\alpha+1} = A_\alpha + C^*_\alpha$ where $|C^*_\alpha| \leq \aleph_0$ and (i) and (ii) further yield that $A = \sum_{\alpha\in I} C^*_\alpha$.

<u>Corollary 66</u>. A direct summand of an Axiom 2 group is an Axiom 2 group.

<u>Theorem 67</u>. A direct summand of an Axiom 3 group is an Axiom 3 group.

<u>Proof</u>. Let $\mathcal{C}$ be a collection of nice subgroups of a group G that satisfies Axiom 3. Suppose that $G = H + K$. Define $\mathcal{C}_H$ to be the collection of subgroups A of H such that $N = A + (N \cap K)$ for some $N \in \mathcal{C}$. Such an A is nice in H, $0 \in \mathcal{C}_H$ and $\mathcal{C}_H$ is closed with respect to ascending unions. Suppose that $A \in \mathcal{C}_H$. Then $N = A + (N \cap K)$ where $A \subseteq H$ and N is an element of $\mathcal{C}$. Let B be a subgroup of H such that $\{A,B\}/A$ is countable. Then $\{N,B\}/N$ is countable and there exists $M_1 \in \mathcal{C}$ such that $M_1 \supseteq \{N,B\}$ and such that M_1/N is countable. Since $|M_1/N| \leq \aleph_0$ and since $N = A + N \cap K$ there exists subgroups H_1 and K_1 of H and K that are countable extensions of A and $N \cap K$, respectively, such that $M_1 \subseteq H_1 + K_1$. In turn, there exists M_2 in $\mathcal{C}$ such that $M_2 \supseteq \{N,H_1\}$ and such that M_2/N is countable. Continuing in this way, we obtain sequences $M_i \in \mathcal{C}$, $H_i \subseteq H$, and $K_i \subseteq K$ with the property that $M_i \subseteq H_i + K_i$, $M_{i+1} \supseteq \{N,H_i\}$ and $|M_i/N| \leq \aleph_0$

for each positive integer i. Setting $M = \bigcup_{i<\omega} M_i$, we have that $M = (M \cap H) + (M \cap K)$. Thus, $M \cap H \in \mathcal{C}_H$. Since $M \cap H \supseteq \{A,B\}$ and since $|(M \cap H)/A| \leq |M/N| \leq \aleph_0$, the theorem is proved.

Theorem 68. Let β denote an arbitrary ordinal. If $G/p^{\beta}G$ is an Axiom 3 group and if $p^{\beta+1}G = 0$, then G is also an Axiom 3 group.

Proof. Let $\mathcal{C}'$ be a collection of nice subgroups of $G/p^{\beta}G$ that satisfies the conditions of Axiom 3. Define $\mathcal{C}$ to be the collection of all subgroups N of G such that $\{N,p^{\beta}G\}/p^{\beta}G \in \mathcal{C}'$. Since $p(p^{\beta}G) = 0$, every subgroup of $p^{\beta}G$ is a direct summand of $p^{\beta}G$; in particular, every subgroup of $p^{\beta}G$ is a nice subgroup of $p^{\beta}G$. Thus, each member of $\mathcal{C}$ is a nice subgroup of G by Proposition 60. Clearly, $0 \in \mathcal{C}$ and $\mathcal{C}$ is closed with respect to ascending unions. Suppose that $N \in \mathcal{C}$ and that H is a subgroup of G such that $\{H,N\}/N$ is countable. Let $H = \{H \cap N, C\}$ where C is countable. Since $N \in \mathcal{C}$, $\{N,p^{\beta}G\}/p^{\beta}G \in \mathcal{C}'$. There exists $M/p^{\beta}G \in \mathcal{C}'$ such that $M \supseteq \{N,H\}$ and such that $|M/\{N,p^{\beta}G\}| \leq \aleph_0$, for $\{\{N,p^{\beta}G\},\{H,p^{\beta}G\}\}/p^{\beta}G/\{N,p^{\beta}G\}/p^{\beta}G$ is countable. Let $M = \{N,C',p^{\beta}G\}$ where C' countable. Define $M_0 = \{N,C,C'\}$. Then $\{M_0,p^{\beta}G\}/p^{\beta}G = M/p^{\beta}G$ is in $\mathcal{C}'$, so M_0 is in $\mathcal{C}$. Obviously, $M_0 \supseteq \{N,H\}$ and M_0/N is countable. This shows that G satisfies Axiom 3.

Our next project having established the above properties of Axiom 3 groups is to construct a class of "cannonical" Axiom 3 groups. However, this class of p-groups overlaps some of the material in the succeeding chapter and so we briefly introduce a few notions needed in that chapter.

Definition. A functor S on the category $\mathcal{A}$ of abelian groups is called a preradical if, for each group A in $\mathcal{A}$, $SA \subseteq A$ and if $f: A \to B$, then $S(f) = f|SA$. If, in addition, $S(A/SA) = 0$ for all A in $\mathcal{A}$, we call S a radical on $\mathcal{A}$. A short exact sequence $A \rightarrowtail B \twoheadrightarrow C$ is called S-pure if it represents an element of SExt(C,A). If S is a radical having a representing sequence in the sense that there is an exact sequence $Z \rightarrowtail M \twoheadrightarrow H$ such that for any A in $\mathcal{A}$ we have that SA = Image(Hom(M,A) → Hom(Z,A) ≃ A), then S is called a cotorsion functor.

<u>Examples</u>. (1) $SG = G[p]$ is a preradical but not a radical.

(2) $SG = tG_p$ is a radical but not a cotorsion functor (verification is left to the reader).

(3) $SG = p^\alpha G$ (for a fixed prime p and ordinal α) is a cotorsion functor as we show below.

<u>Theorem 69</u>. For each ordinal $\alpha \geq 0$, there is a p-group H_α and an extension $e_\alpha\colon Z \rightarrowtail M_\alpha \twoheadrightarrow H_\alpha$ which represents the radical p^α and

(i) $H_\alpha = \sum_{\beta<\alpha} H_\beta$ if α is a limit ordinal.

(ii) $p^\alpha H_{\alpha+1}$ is cyclic of order p and $H_\alpha \simeq H_{\alpha+1}/p^\alpha H_{\alpha+1}$.

(iii) length $H_\alpha = \alpha$.

(iv) $1 \leq f_{H_\alpha}(\beta) \leq |\alpha|$ for each $\beta < \alpha$, when $\alpha \geq \omega$.

(v) H_α is an Axiom 3 group for each α.

<u>Proof</u>. For $\alpha = 0$, take $H_0 = 0$ and $e_0\colon Z \rightarrowtail Z \twoheadrightarrow 0$. For $\alpha = 1$, take $H_1 = Z(p)$ and $e_1\colon Z \overset{p}{\rightarrowtail} Z \twoheadrightarrow Z(p)$. In either case (i)-(v) are clearly satisfied. Now suppose that the collection $[H_\gamma]_{\gamma<\alpha}$ together with the extensions $[e_\gamma\colon Z \rightarrowtail M_\gamma \twoheadrightarrow H_\gamma]_{\gamma<\alpha}$ have been constructed so as to satisfy the above conditions. If α is a limit ordinal, we choose an extension $e_\alpha\colon Z \rightarrowtail M_\alpha \twoheadrightarrow \sum_{\beta<\alpha} H_\beta$ which represents the element $\langle e_\beta \rangle_{\beta<\alpha}$ in $\operatorname{Ext}(\sum_{\beta<\alpha} H_\beta, Z) = \prod_{\beta<\alpha} \operatorname{Ext}(H_\beta, Z)$, where e_β is represented by $Z \rightarrowtail M_\beta \twoheadrightarrow H_\beta$ for $\beta < \alpha$. Take $H_\alpha = \sum_{\beta<\alpha} H_\beta$ and observe that (i)-(v) are easily verified. It remains to show that p^α is represented by $e_\alpha\colon Z \rightarrowtail M_\alpha \twoheadrightarrow H_\alpha$. For each $f\colon Z \to X$ which extends to $\bar{f}\colon M \to X$ and for each ordinal $\beta < \alpha$, there is a commutative diagram

(*)

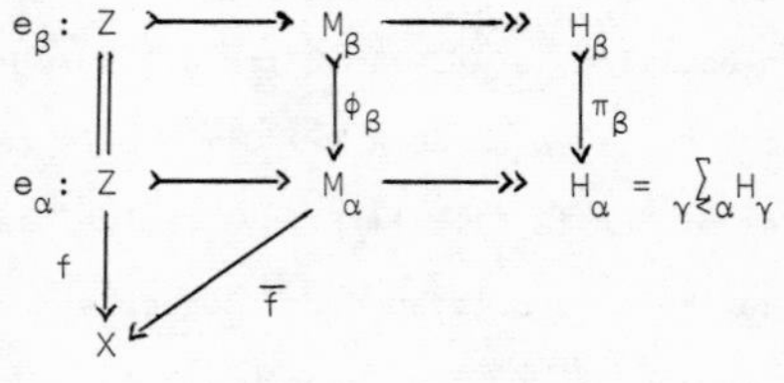

where π_β is the β-th coordinate injection of H_β into H_α (resulting from

$e_\alpha = \langle e_\beta \rangle_{\beta<\alpha}$). It follows that if $x \in \mathrm{Image}(\mathrm{Hom}(M_\alpha,X) \to \mathrm{Hom}(Z,X) \simeq X)$, then $x \in p^\beta X$ for all $\beta < \alpha$ which implies that $\mathrm{Image}(\mathrm{Hom}(M_\alpha,X) \to X) \subseteq p^\alpha X$. If, on the other hand, $x \in p^\alpha X$, then $x \in p^\beta X$ for each $\beta < \alpha$ and, hence, there is a map $f_\beta\colon M_\beta \to X$, $\beta < \alpha$, such that $f_\beta(n) = nx$ for each $n \in Z$. By identifying M_β with its image in the commutative diagram (*), we obtain an exact sequence $K \rightarrowtail \sum_{\beta<\alpha} M_\beta \overset{\sigma}{\twoheadrightarrow} M_\alpha$ where $\sigma(y_{\beta_1},\ldots,y_{\beta_n}) = y_{\beta_1} + \ldots + y_{\beta_n} \in M_\alpha$. Define $f\colon \sum_{\beta<\alpha} M_\beta \to X$ by $f(y_{\beta_1},\ldots,y_{\beta_n}) = f_{\beta_1}(y_{\beta_1}) + \ldots + f_{\beta_n}(y_{\beta_n})$. If $(y_{\beta_1},\ldots,y_{\beta_n}) \in \mathrm{Ker}\sigma$, then $y_{\beta_1} + \ldots + y_{\beta_n} = 0$ which implies that $\sum_{i=1}^{n} (y_{\beta_i} + Z) = 0$ and, therefore, that $y_{\beta_i} = m_i \in Z$ for $i = 1,\ldots,n$. Since $f(y_{\beta_1} + \ldots + y_{\beta_n}) = f(m_1,\ldots,m_n) = f_{\beta_1}(m_1) + \ldots + f_{\beta_n}(m_n) = m_1x + \ldots + m_nx = (m_1 + \ldots + m_n)x = 0\cdot x = 0$, it follows that $f(\mathrm{Ker}\sigma) = 0$. Thus, f induces a map $f^*\colon M_\alpha \to X$ such that $f^*(1) = x$. This shows that $p^\alpha X \subseteq \mathrm{Image}(\mathrm{Hom}(M_\alpha,X) \to X)$. Thus, $p^\alpha X = \mathrm{Image}\,(\mathrm{Hom}(M_\alpha,X) \to X)$ yielding that e_α represents p^α.

Suppose that $\alpha = \beta + 1$ and define $H_{\beta+1} = M_\beta/pZ$. This yields a commutative diagram

$$\begin{array}{ccccc}
Z & \overset{p}{\rightarrowtail} & M_{\beta+1} & \twoheadrightarrow & H_{\beta+1} \\
{\scriptstyle p}\downarrow & & \| & & \downarrow{\scriptstyle \theta} \\
Z & \rightarrowtail & M_\beta & \twoheadrightarrow & H_\beta
\end{array}$$

where $\mathrm{Ker}\sigma = Z/pZ \simeq Z(p)$. Now (i)-(iv) are elementary to prove while (v) follows from Theorem 68 for the collection $[H_\gamma]_{\gamma\le\alpha}$. Furthermore, it is easily seen from the above diagram that $\mathrm{Image}(\mathrm{Hom}(M_{\beta+1},X) \overset{p}{\to} \mathrm{Hom}(Z,X) \simeq X) = p\mathrm{Image}(\mathrm{Hom}(M_\beta,X) \to X)$, that is, $\mathrm{Image}(\mathrm{Hom}(M_\beta,X) \overset{p}{\to} X) = p(p^\beta X) = p^{\beta+1}X = p^\alpha X$. This completes our proof.

We shall refer to the group H_α above as the <u>generalized Prüfer group of length α</u>, since $H_{\omega+1}$ is the so-called Prüfer group defined in Exercise 10. We let $\mathcal{H}$ be the class of all p-primary groups which are isomorphic to direct summands of direct sums of copies of generalized Prüfer groups.

<u>Theorem 70</u>. (Hill) If $\mathcal{K}$ is a class of reduced p-groups which contains

the class $\mathcal{H}$, is closed with respect to taking direct sums and the Ulm invariants distinguish between nonisomorphic groups in $\mathcal{K}$, then $\mathcal{K} = \mathcal{H}$.

Proof. Let G be in $\mathcal{K}$ of length α and let $m = |G|\aleph_0$. Let $H = \sum_{\beta \leq \alpha} \sum_m H_\beta$ and observe via Theorem 69(iv) that G + H and H have the same Ulm invariants. Hence, $G + H \simeq H$. Thus, G must be in $\mathcal{H}$ which completes the proof.

Corollary 71. $\mathcal{H}$ is precisely the class of Axiom 3 groups.

Corollary (to proof) 72. G is an Axiom 3 group of length $\leq \Omega$ if and only if G is a reduced d.s.c. p-group.

Corollary 73. If G is an Axiom 3 group of limit length β, then G is a direct summand of a direct sum of Axiom 3 groups of lengths less than β.

The following two results were first proved by Nunke [105] for the class of totally projective p-groups. In the following chapter we shall show that this class coincides with the class of Axiom 3 groups.

Theorem 74. (Nunke) Let G be a p-group and let β be an arbitrary ordinal. Then $p^\beta G$ and $G/p^\beta G$ are Axiom 3 groups if and only if G is an Axiom 3 group.

Proof. First suppose that G is an Axiom 3 group and let $\mathcal{C}$ be a collection of nice subgroups satisfying Axiom 3. Define $\mathcal{C}_{p^\beta G} = [N \cap p^\beta G \colon N \in \mathcal{C}]$ and $\mathcal{C}_{G/p^\beta G} = [\{N, p^\beta G\}/p^\beta G \colon N \in \mathcal{C}]$. By Proposition 60, $\mathcal{C}_{p^\beta G}$ and $\mathcal{C}_{G/p^\beta G}$ are collections of nice subgroups of $p^\beta G$ and $G/p^\beta G$, respectively. Clearly, both collections satisfy the conditions of Axiom 3 and, hence, $p^\beta G$ and $G/p^\beta G$ are Axiom 3 groups.

Now suppose that $G/p^\beta G$ and $p^\beta G$ are Axiom 3 groups. Let $m = |G|\aleph_0$, let α be the length of G and let $H = \sum_{\beta \leq \alpha} \sum_m H_\beta$. As in the proof of Theorem 70, $p^\beta H$ and $p^\beta(G + H)$ have the same Ulm invariants as also do G and H. The proof of the sufficiency yields that $p^\beta H$ and $p^\beta(G + H) = p^\beta G + p^\beta H$ are Axiom 3 groups. Therefore, there is an isomorphism $\phi\colon p^\beta H \rightarrowtail\!\!\!\!\rightarrow p^\beta(G + H)$ and, of course, ϕ is height-preserving. By Exercise 5, $F_\alpha(H, p^\beta H) = F_\alpha(G + H, p^\beta(G + H))$ for all α. Thus, we may apply Theorem 63 and extend ϕ to an isomorphism $\bar{\phi}\colon H \rightarrowtail\!\!\!\!\rightarrow G + H$. It follows that G is an Axiom 3 group.

Corollary 75. (Nunke) Let G be a reduced p-group and let $\alpha \leq \Omega$. Then $G/p^{\alpha}G$ and $p^{\alpha}G$ are d.s.c. groups if and only if G is a d.s.c. group.

In closing this chapter on the classification of groups according to cardinal invariants, we mention a criterion suggested by Kaplansky [69] that one might apply in deciding whether or not a set of invariants is useful. This criterion is in the form of two test problems.

Problem I. If G is isomorphic to a direct summand of H and if H is isomorphic to a direct summand of G, are G and H isomorphic?

Problem II. If G + G and H + H are isomorphic, are G and H isomorphic?

These test problems have affirmative answers for Axiom 3 groups (see Exercise 13).

Exercises

1. Let $B = \sum_{n<\omega} Z(p^n)$ and let $\overline{B}$ be the torsion completion of B (see Exercise 14, Chapter III). Prove that $f_B(n) = f_{\overline{B}}(n) = 1$ for $n < \omega$ and that $f_B(\alpha) = f_{\overline{B}}(\alpha) = 0$ for $\alpha \geq \omega$. Prove a more general result concerning $f_G(\alpha)$ and $f_B(\alpha)$ where G is a reduced p-group and B is a basic subgroup.

2. Let $A = \sum_i A_i$ (p-groups) and show that $f_A(\alpha) = \sum_i f_{A_i}(\alpha)$ for each α.

3. Let $G = B + \sum_{n<\omega} \sum_{2^{\aleph_0}} Z(p^n)$ and let $H = \overline{B} + \sum_{n<\omega} \sum_{2^{\aleph_0}} Z(p^n)$. Then $|G| = |H| = 2^{\aleph_0}$. Prove that $H \neq G$ but $f_G(\alpha) = f_H(\alpha)$ for each α.

4. In case A is a direct summand of G, show that $F_{\alpha}(G,A) = F_{\alpha}(G/A,0) = f_{G/A}(\alpha)$ for each α.

5. $F_{\alpha}(G,p^{\beta}G) = F_{\alpha}(G,0) = f_G(\alpha)$ for $\alpha < \beta$ and $F_{\alpha}(G,p^{\beta}G) = 0$ for $\alpha \geq \beta$. If β is a limit ordinal, then $F_{\alpha}(G,p^{\beta}G) = F_{\alpha}(G/p^{\beta}G,0) = f_{G/p^{\beta}G}(\alpha)$, for $\alpha < \beta$.

6. Prove Proposition 54.

7. Prove Proposition 58.

8. Prove Proposition 59.

9. Prove the second half of Proposition 60.

10. Let $A \simeq Z(p^{\infty})$ and $B = \sum_{n<\omega} \{b_n\}$ where $\{b_n\} \simeq Z(p^n)$ for each n. Let $a_o \neq 0 \in A[p]$ and choose $a_n \in A$ such that $p^n a_n = a_o$ for $n < \omega$. Set $g_n = a_n + b_n \in A + B$ for $n < \omega$ and let G be the subgroup of A + B generated by $[g_n]_{n<\omega}$. Prove that

(i) G is reduced;

(ii) $p^{\omega}G = \{a_o\}$ and $G/p^{\omega}G \simeq B$;

(iii) $G \simeq H_{\omega+1}$.

G has often been referred to as the Prüfer group.

11. Let G be a reduced p-group and suppose that H is a subgroup of G such that G/H is a direct sum of cyclic groups. Show that every height-preserving automorphism of H extends to an automorphism of G.

13. Show that if G and H are as in Kaplansky's Test Problems I or II, then G and H have the same Ulm invariants. Hence, I and II have affirmative answers for Axiom 3 groups.

14. Let G be a reduced p-group and let $x \in G$. Define the <u>Ulm sequence of x in G</u> by $U^G(x) = (\beta_1, \beta_2, \ldots, \beta_n, \ldots)$ where $\beta_n = h^G(p^n x)$. G is called <u>transitive</u> if, for each $x, y \in G$ with $U^G(x) = U^G(y)$, there is an automorphism ϕ of G such that $\phi(x) = y$. Use Theorem 63 to show that an Axiom 3 group is transitive.

15. Using the notation of Exercise 14, we say that a p-group G is <u>fully transitive</u> if, for each $x, y \in G$ with $U^G(x) \leq U^G(y)$, there is an endomorphism ϕ of G such that $\phi(x) = y$. Modify the proof of Theorem 63 to show that any Axiom 3 group is fully transitive.

16. Prove, for any ordinal α, that $p^{\beta}(G/p^{\alpha}G) = p^{\beta}G/p^{\alpha}G$ for each $\beta \leq \alpha$.

17. In Theorem 69, show that $H_n \simeq Z(p^n)$ for $n < \omega$.

VI. THE TOTALLY PROJECTIVE GROUPS AND THE FUNCTORS p^{α}

There is an attempt made in this chapter to explore lightly some of the more "homological" aspects in abelian group theory which are mainly due to Nunke [103], [104], [105]. At the same time, we obtain some significant structural results and the existence of p-groups with rather varied properties. Preceding our main theme in this chapter are two special cases of theorems quoted from Nunke [102] on homological algebra. The proofs of these results have been set aside for now and can be found in the Appendix. The maps in the following two theorems are natural and, hence, generally follow the route of pushout or pullback diagrams. Furthermore, most of the homological results that follow are simply consequences of these two theorems and our elementary knowledge of Hom, Ext, $\otimes$, and Tor.

<u>Theorem 76</u>. (Nunke) If $Z \rightarrowtail M \twoheadrightarrow H$ is an exact sequence inducing a connecting homomorphism ∂: $\mathrm{Tor}(H,B) \to Z \otimes B \cong B$, then there is a commutative diagram

$$\begin{array}{ccc} \mathrm{Ext}(B,C) \cong \mathrm{Ext}(Z \otimes B,C) & \xrightarrow{\mathrm{Ext}(\partial,C)} & \mathrm{Ext}(\mathrm{Tor}(H,B),C) \\ & \searrow^{\delta} & \downarrow \mu \\ & & \mathrm{Ext}(H,\mathrm{Ext}(B,C)) \end{array}$$

where $\mathrm{Ext}(\partial,C)$ is defined by the pullback diagram

$$\begin{array}{rccccc} \mathrm{Ext}(\partial,C)(e): & C & \rightarrowtail & Y & \twoheadrightarrow & \mathrm{Tor}(H,B) \\ & \| & & \downarrow & & \downarrow \partial \\ e: & C & \rightarrowtail & X & \twoheadrightarrow & B \end{array}$$

and δ is the connecting homomorphism δ_E: $E \cong \mathrm{Hom}(Z,E) \to \mathrm{Ext}(H,E)$ induced by the exact sequence $Z \rightarrowtail M \twoheadrightarrow H$ with $E = \mathrm{Ext}(B,C)$.

<u>Theorem 77</u>. (Nunke) If $Z \rightarrowtail M \twoheadrightarrow H$ is an exact sequence inducing a

connecting homomorphism $\delta: C \cong \mathrm{Hom}(Z,C) \to \mathrm{Ext}(H,C)$, then there is a commutative diagram

$$\begin{array}{ccc} \mathrm{Ext}(A,C) \cong \mathrm{Ext}(A,\mathrm{Hom}(Z,C)) & \xrightarrow{\mathrm{Ext}(A,\delta)} & \mathrm{Ext}(A,\mathrm{Ext}(H,C)) \\ & \searrow^{\Delta} & \downarrow \\ & & \mathrm{Ext}(H,\mathrm{Ext}(A,C)) \end{array}$$

where $\mathrm{Ext}(A,\delta)$ is defined by the pushout diagram

$$\begin{array}{rcccc} e: C & \rightarrowtail & X & \twoheadrightarrow & A \\ \delta \downarrow & & \downarrow & & \Vert \\ \mathrm{Ext}(A,\delta)(e): \mathrm{Ext}(H,C) & \rightarrowtail & Y & \twoheadrightarrow & A \end{array}$$

and Δ is the connecting homomorphism $\Delta_E: E = \mathrm{Hom}(Z,E) \to \mathrm{Ext}(H,E)$ induced by the exact sequence $Z \rightarrowtail M \twoheadrightarrow H$ with $E = \mathrm{Ext}(A,C)$.

Corollary 78. For any groups A, B and C, we have a natural homomorphism $\mathrm{Ext}(\mathrm{Tor}(A,B),C) \to \mathrm{Ext}(A,\mathrm{Ext}(B,C))$.

The above theorems are instrumental in studying the p^α functors introduced near the end of the last chapter (see the discussion preceding Theorem 69).

Definition. A short exact sequence $A \rightarrowtail B \twoheadrightarrow C$ is called p^α-pure or A is called a p^α-pure subgroup of B if this extension represents an element of $p^\alpha\mathrm{Ext}(C,A)$.[2] A group G is called p^α-projective (p^α-injective) if $p^\alpha\mathrm{Ext}(G,X) = 0$ ($p^\alpha\mathrm{Ext}(X,G) = 0$) for all groups X. Moreover, a reduced p-group G is called totally projective if $p^\alpha\mathrm{Ext}(G/p^\alpha G,X) = 0$ for all ordinals α and all groups X.

Lemma 79. If G is a reduced p-group such that $p^{\alpha+1}G = 0$ and $G/p^\alpha G$ is totally projective, then G is totally projective.

Proof. First suppose that $A \xrightarrow{\delta} B \xrightarrow{\pi} C \xrightarrow{\sigma} D$ is exact with $pA = 0 = pD$ and with $p^\alpha B = 0$. If $\alpha < \omega$, then clearly $p^{\alpha+1}C = 0$. If $\alpha \geq \omega$,

[2] Our definition of "p-pure" in Chapter II corresponds, it turns out, to "p^ω-pure" in this chapter.

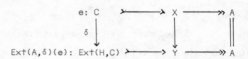

then the exactness of $B[p]/\text{Image} \rightarrowtail B/\text{Image}\delta \twoheadrightarrow pB$ and the fact that $p^{\alpha}(pB) = p^{\alpha}B = 0$ together imply that $p^{\alpha+1}(B/\text{Image}\delta) = 0$, that is, $p^{\alpha+1}(\text{Image}\pi) = 0$. Since clearly $pC \subseteq \text{Image}\pi \subseteq C$, it follows that $p^{\beta}C = p^{\beta}(\text{Image}\pi)$ for $\beta \geq \omega$. Hence, $p^{\alpha+1}C = p^{\alpha+1}(\text{Image}\pi) = 0$.

From the exact sequence

$$\text{Hom}(p^{\alpha}G,X) \xrightarrow{\delta} \text{Ext}(G/p^{\alpha}G,X) \longrightarrow \text{Ext}(G,X) \twoheadrightarrow \text{Ext}(p^{\alpha}G,X),$$

we obtain an exact sequence as above with $A = \text{Hom}(p^{\alpha}G,X)$, $B = \text{Ext}(G/p^{\alpha}G,X)$, $C = \text{Ext}(G,X)$ and $D = \text{Ext}(p^{\alpha}G,X)$. Thus, $p^{\alpha+1}\text{Ext}(G,X) = 0$ for all X. This shows that G is totally projective, since $G/p^{\beta}G \simeq G/p^{\alpha}G/p^{\beta}G/p^{\alpha}G \simeq (G/p^{\alpha}G)/p^{\beta}(G/p^{\alpha}G)$ is already p^{β}-projective for $\beta \leq \alpha$.

Theorem 80. The generalized Prüfer group H_{α} is totally projective for each α.

Proof. The proof is accomplished by induction on the length of $H_{\alpha} = \alpha$. For $\alpha = 0$, the result is trivial. If α is a limit ordinal, then $H_{\alpha} = \sum_{\beta<\alpha} H_{\beta}$. By our induction hypothesis, H_{β} is totally projective for each $\beta < \alpha$ and so H_{α} is also totally projective (direct sums of totally projectives are totally projective). If $\alpha = \beta + 1$, then by Theorem 69(ii) $H_{\beta+1}/p^{\beta}H_{\beta+1} \simeq H_{\beta}$. Thus, our induction hypothesis and an application of Lemma 79 establishes that $H_{\beta+1}$ is totally projective.

Corollary 81. The generalized Prüfer group H_{α} is p^{α}-projective for each ordinal α.

Remark. The representing sequence $e_{\alpha}: Z \rightarrowtail M_{\alpha} \twoheadrightarrow H_{\alpha}$ yields for any group G an exact sequence

$$p^{\alpha}G \rightarrowtail G \xrightarrow{\delta_G} \text{Ext}(H_{\alpha},G).$$

We shall make a good deal of use of this exact sequence in the next several theorems. For example, this sequence is used in the proof of Theorem 84 (below) with $G = \text{Ext}(A,K)$.

Theorem 82. If a group A is p^{α}-projective, then $\text{Tor}(A,X)$ and $\text{Ext}(A,Y)$ are, respectively, p^{α}-projective and p^{α}-injective for any groups X and Y.

Proof. For any groups X and Y, we have the isomorphism (Corollary 78)

$$0 = p^{\alpha}\mathrm{Ext}(A,\mathrm{Ext}(X,Y)) \simeq p^{\alpha}\mathrm{Ext}(\mathrm{Tor}(A,X),Y) \simeq p^{\alpha}\mathrm{Ext}(X,\mathrm{Ext}(A,Y)),$$

which proves our assertion.

<u>Corollary 83</u>. For a generalized Prüfer group H_{α}, $\mathrm{Tor}(H_{\alpha},X)$ is p^{α}-projective and $\mathrm{Ext}(H_{\alpha},X)$ is p^{α}-injective for any group X.

The next result establishes that there are "enough" p^{α}-projectives and p^{α}-injectives.

<u>Theorem 84</u>. Given a group A, there are p^{α}-pure projective and p^{α}-pure injective resolutions, respectively:

(a) $K \rightarrowtail F + \mathrm{Tor}(H_{\alpha},A) \xrightarrow{\sigma_A} A$ with F free and $\sigma_A = \theta + \partial_A$, where ∂_A: $\mathrm{Tor}(H_{\alpha},A) \to Z \otimes A \simeq A$ is the connecting homomorphism and θ is any homomorphism of F into A such that $\theta + \partial_A$ is an epimorphism.

(b) $A \xrightarrow{\pi_A} \mathrm{Ext}(H_{\alpha},A) + D(p^{\alpha}A) \twoheadrightarrow U$ with $\pi_A = \delta_A + \phi$, where δ_A: $A \simeq \mathrm{Hom}(Z,A) \to \mathrm{Ext}(H_{\alpha},A)$ is the connecting homomorphism and ϕ: $A \to D(p^{\alpha}A)$ is any homomorphism that extends the embedding of $p^{\alpha}A$ into $D(p^{\alpha}A)$.

<u>Proof</u>. By Theorem 82 and Corollary 83, $F + \mathrm{Tor}(H_{\alpha},A)$ is p^{α}-projective. Using the representing sequence $Z \rightarrowtail M_{\alpha} \twoheadrightarrow H_{\alpha}$ for p^{α}, we obtain an exact sequence $p^{\alpha}\mathrm{Ext}(A,K) \rightarrowtail \mathrm{Ext}(A,K) \xrightarrow{\delta} \mathrm{Ext}(H_{\alpha},\mathrm{Ext}(A,K))$. Applying Theorem 76, we obtain the commutative diagram

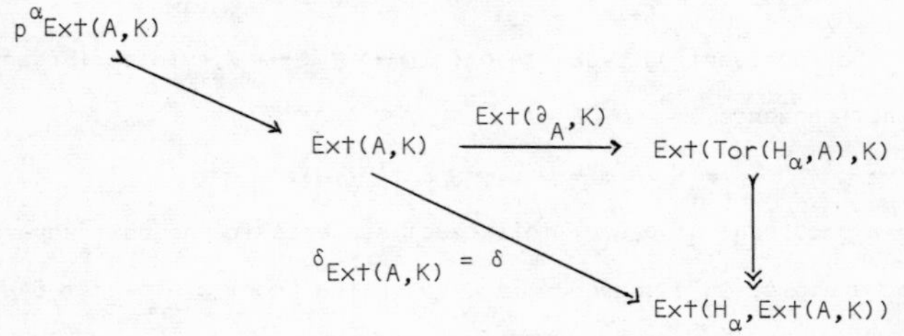

Let e denote the element of Ext(A,K) represented by $K \rightarrowtail F + \mathrm{Tor}(H_{\alpha},A) \xrightarrow{\sigma_A} A$. To show that $e \in p^{\alpha}\mathrm{Ext}(A,K) = \mathrm{Ker}\,\delta$, observe from the above diagram that this is equivalent to showing that

$Ext(\partial_A, K)(e) = 0$. Since $\sigma_A = \theta + \partial_A$, there is a pullback diagram

$$
\begin{array}{ccccc}
e': K & \rightarrowtail & K + Tor(H_\alpha, A) & \xrightarrow{\rho} & Tor(H_\alpha, A) \\
& \| & \downarrow{\psi} & & \downarrow{\partial_A} \\
e: K & \rightarrowtail & F + Tor(H_\alpha, A) & \xrightarrow{\sigma_A} & A
\end{array}
$$

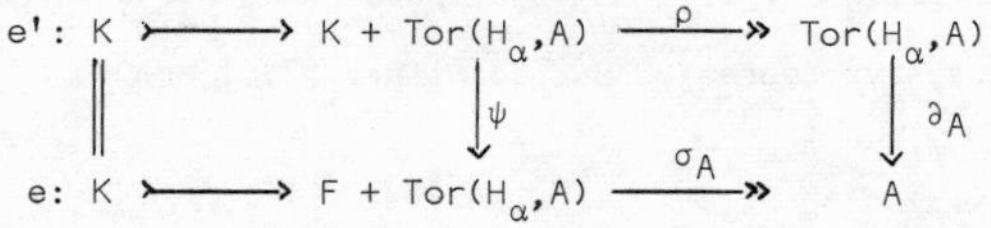

where ρ is the natural projection of $K + Tor(H_\alpha, A)$ onto $Tor(H_\alpha, A)$ and $\psi(k, t) = k + \partial_A(t)$ for $k \in K$ and $t \in Tor(H_\alpha, A)$. By definition (see Theorem 76), $Ext(\partial_A, K)(e) = e' = 0$, since e' is represented by a split exact sequence. The proof of (b) is dual to the above and makes use of Theorem 77. Its proof is left as an exercise (see Exercise 23).

An important observation at this point is that a proof of the following lemma is embodied in the above proof of Theorem 84.

<u>Lemma 85.</u> The extension $e: K \overset{i}{\rightarrowtail} H \overset{j}{\twoheadrightarrow} G$ is p^α-pure if and only if either (a) $Ext(\partial_G, K)(e) = 0$, where $Ext(\partial_G, K): Ext(G, K) \rightarrow Ext(Tor(H_\alpha, G), K)$

or equivalently there is a commutative diagram

or (b) $Ext(G, \delta_K)(e) = 0$, where $Ext(G, \delta_K): Ext(G, K) \rightarrow Ext(G, Ext(H_\alpha, K))$

or equivalently there is a commutative diagram

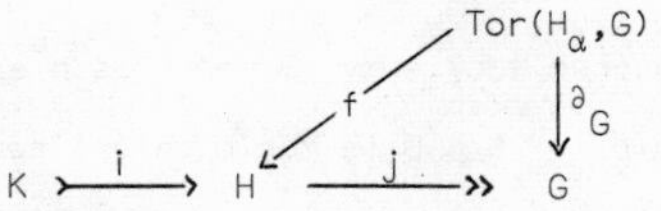

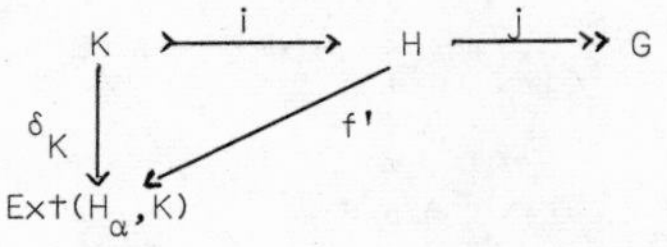

Of course, the proof of part (b) in the above lemma is contained in the proof (left to reader) of part (b) in Theorem 84. In the next few results we are concerned with other necessary and/or sufficient conditions for an extension to be p^α-pure.

<u>Proposition 86.</u> If the extension $A \overset{i}{\rightarrowtail} B \twoheadrightarrow C$ is p^α-pure, then

$A \cap p^{\alpha}B = p^{\alpha}A$, that is, the induced homomorphism $A/p^{\alpha}A \to B/p^{\alpha}B$ is monic.

<u>Proof</u>. Since $A \rightarrowtail B \twoheadrightarrow C$ is p^{α}-pure and since $\mathrm{Ext}(H_{\alpha},A) + D(p^{\alpha}A)$ is p^{α}-injective, by Theorem 82 and Corollary 83, there is a commutative diagram

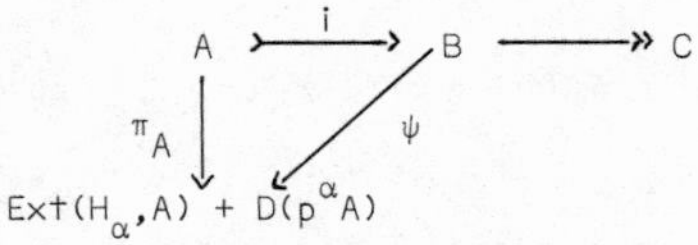

where π_A is as in Theorem 84(b). Therefore, $\pi_A(p^{\alpha}A) \subseteq \psi(p^{\alpha}B) \subseteq D(p^{\alpha}A)$, since $p^{\alpha}\mathrm{Ext}(H_{\alpha},A) = 0$. If $a \in A \cap p^{\alpha}B$, then $\pi_A(a) = \psi(a) \in D(p^{\alpha}A)$. But by definition of π_A, it follows that $\pi_A(a) = \delta_A(a) + \phi(a) = \phi(a)$, that is, $\delta_A(a) = 0$. Hence, $a \in \mathrm{Ker}\delta_A = p^{\alpha}A$ and so $A \cap p^{\alpha}B \subseteq p^{\alpha}A$. The other inclusion being obvious, we have shown that $A \cap p^{\alpha}B = p^{\alpha}A$.

<u>Proposition 87</u>. The extension $A \rightarrowtail B \xrightarrow{\nu} C$ is p^{α}-pure if and only if the induced map $A/p^{\alpha}A \to B/p^{\alpha}B$ is monic and $A/p^{\alpha}A \rightarrowtail B/p^{\alpha}B \xrightarrow{\nu^*} C^*$ is p^{α}-pure, where $C^* = C/\nu(p^{\alpha}B)$ and ν^* is induced by ν.

<u>Proof</u>. First suppose that $A \rightarrowtail B \xrightarrow{\nu} C$ is p^{α}-pure. By Proposition 86, the induced map $A/p^{\alpha}A \to B/p^{\alpha}B$ is monic which induces a short exact sequence e^*: $A/p^{\alpha}A \xrightarrow{\rho} B/p^{\alpha}B \xrightarrow{\nu^*} C^*$. Since $A \rightarrowtail B \twoheadrightarrow C$ is p^{α}-pure exact and $\mathrm{Ext}(H_{\alpha},A/p^{\alpha}A)$ is p^{α}-injective (Corollary 83), we have a commutative diagram

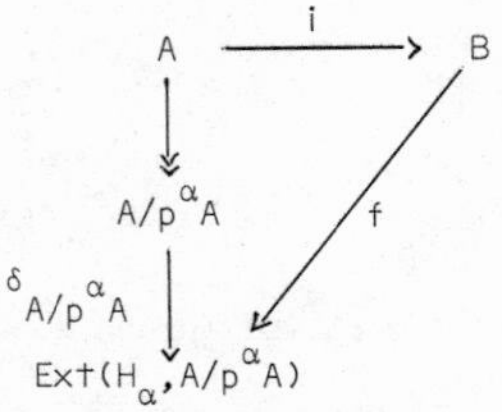

and, moreover, $f(p^{\alpha}B) = 0$, since $p^{\alpha}\mathrm{Ext}(H_{\alpha},A/p^{\alpha}A) = 0$ by Corollary 81. Hence, f induces a map f^*: $B/p^{\alpha}B \to \mathrm{Ext}(H_{\alpha},A/p^{\alpha}A)$ and from the above diagram it is elementary to see that $f^*\rho = \delta_{A/p^{\alpha}A}$. By Lemma 85(b),

$e^* \in p^{\alpha} \text{Ext}(C^*, A/p^{\alpha}A)$.

If ρ is monic and $e^*: A/p^{\alpha}A \rightarrowtail B/p^{\alpha}B \xrightarrow{\nu^*} C^*$ is p^{α}-pure, we use the exact sequence $p^{\alpha}A \rightarrowtail A \xrightarrow{\delta_A} \text{Ext}(H_{\alpha}, A)$ and the p^{α}-injectivity of $\text{Ext}(H_{\alpha}, A)$ to obtain the bottom triangle in the commutative diagram

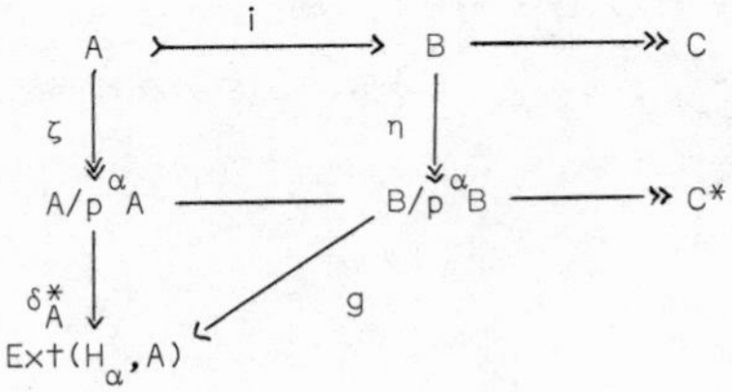

where δ^*_A is induced by δ_A. (Remember $\delta_A(p^{\alpha}A) = 0$!) Hence, $\delta_A = \delta^*_A \zeta$ and $f = g\eta$. It follows that $fi = \delta_A$. By Lemma 85(b), $A \rightarrowtail B \twoheadrightarrow C$ is p^{α}-pure.

For $\alpha \leq \omega$, one can prove a stronger result which we have delegated as an exercise (Exercise 10).

<u>Theorem 88.</u> For $\alpha \leq \omega$, $A \rightarrowtail B \twoheadrightarrow C$ is p^{α}-pure if and only if $A \cap p^{\beta}B = p^{\beta}A$ for all $\beta \leq \alpha$. Hence, A is a pure subgroup of B if and only if $A \rightarrowtail B \twoheadrightarrow B/A$ represents an element of $\bigcap_{n<\omega} n\text{Ext}(B/A, A) \equiv \text{Pext}(B/A, A)$.

<u>Theorem 89.</u> If $A \xrightarrow{i} B \xrightarrow{j} C$ is p^{α}-pure exact and if X is any group, then we have the following exact sequences

(a) $...\text{Hom}(X, C) \xrightarrow{\delta} p^{\alpha}\text{Ext}(X, A) \to p^{\alpha}\text{Ext}(X, B) \to p^{\alpha}\text{Ext}(X, C)$,

and (b) $...\text{Hom}(A, X) \xrightarrow{\delta} p^{\alpha}\text{Ext}(C, X) \to p^{\alpha}\text{Ext}(B, X) \to p^{\alpha}\text{Ext}(A, X)$.

<u>Proof.</u> We only establish (a) since the proof of (b) is dual to that of (a). We already have the exact sequence

(*) $\text{Hom}(X, C) \xrightarrow{\delta} \text{Ext}(X, A) \xrightarrow{i_*} \text{Ext}(X, B) \xrightarrow{j_*} \text{Ext}(X, C)$.

If $f \in \text{Hom}(X, C)$, then $\delta(f)$ is represented by the top row of the pullback diagram

$$
\begin{array}{ccccc}
\delta(f): A & \rightarrowtail & B' & \twoheadrightarrow & X \\
\| & & \downarrow & & \downarrow f \\
A & \rightarrowtail & B & \twoheadrightarrow & C
\end{array}
$$

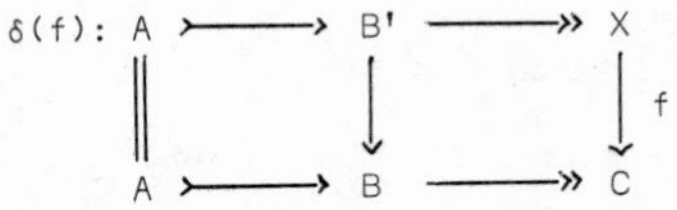

Since the natural map of $Ext(C,A) \to Ext(X,A)$ takes $p^{\alpha}Ext(C,A)$ into $p^{\alpha}Ext(X,A)$, it follows that $\delta(f) \in p^{\alpha}Ext(X,A)$, that is, $\text{Image}\delta \subseteq p^{\alpha}Ext(X,A)$. This fact together with the exactness of (*) yields the sequence

$$Hom(X,C) \xrightarrow{\delta} p^{\alpha}Ext(X,A) \xrightarrow{i_*} p^{\alpha}Ext(X,B) \xrightarrow{j_*} p^{\alpha}Ext(X,A)$$

with exactness at $p^{\alpha}Ext(X,A)$, where i_* and j_* are the restrictions of the identically named maps i_* and j_* in the above exact sequence (*). Hence, it suffices to prove that $\text{Ker}j_* \subseteq \text{Image}i_*$. To prove this fact, we construct the commutative diagram

$$\begin{array}{ccccc}
 & & p^{\alpha}Ext(X,A) & \xrightarrow{i_*} & p^{\alpha}Ext(X,B) \\
 & & \downarrow & & \downarrow \\
Hom(X,C) & \xrightarrow{\delta} & Ext(X,A) & \xrightarrow{i_*} & Ext(X,B) \\
 & & \downarrow{\scriptstyle \delta_{Ext(X,A)} = \delta_1} & & \downarrow{\scriptstyle \delta_{Ext(X,B)} = \delta_2} \\
 & & Ext(H_{\alpha},Ext(X,A)) & \xrightarrow{\theta} & Ext(H_{\alpha},Ext(X,B)) \\
 & & \| & & \| \\
Hom(Tor(H_{\alpha},X),C) & \xrightarrow{\Delta} & Ext(Tor(H_{\alpha},X),A) & \xrightarrow{\phi} & Ext(Tor(H_{\alpha},X),B)
\end{array}$$

where the columns in the top two squares result from the standard exact sequence $p^{\alpha}G \rightarrowtail G \xrightarrow{\delta_G} Ext(H_{\alpha},G)$ and the bottom square is obtained from the natural isomorphism (Corollary 78) $Ext(S,Ext(T,U)) \approx Ext(Tor(S,T),U)$. By the same reasoning as above, it follows that $\text{Image}\Delta \subseteq p^{\alpha}Ext(Tor(H_{\alpha},X),A)$ and, hence, that $\text{Image}\Delta = 0$, since $Tor(H_{\alpha},X)$ is p^{α}-projective. Therefore, both ϕ and θ are monomorphisms. Now suppose that $e \in p^{\alpha}Ext(X,B)$ and that $j_*(e) = 0$. By exactness of the second row in the above diagram, there is $e_1 \in Ext(X,A)$ such that $i_*(e_1) = e$. Hence, $\theta\delta_1(e_1) = \delta_2 i_*(e_1) = \delta_2(e) = 0$, since $e \in p^{\alpha}Ext(X,B)$. This implies that $e_1 \in \text{Ker}\delta_1 = p^{\alpha}Ext(X,A)$, since θ is a monomorphism. This completes our proof.

<u>Lemma 90</u>. Let G be a group and let $x \in p^{\beta}G[p^n]$. Then there is a homomorphism $\phi: H_{\beta+n} \to G$ such that $\phi(b) = x$, where b generates $p^{\beta}H_{\beta+n} = p^{\beta}H_{\beta+n}[p^n]$.

<u>Proof</u>. Our inductive construction of the exact sequences

e_α: $Z \rightarrowtail M_\alpha \twoheadrightarrow H_\alpha$ representing the p^α functors (particularly the case when α is a nonlimit) in the proof of Theorem 69 shows that $M_{\beta+n} = M_\beta$ for n a positive integer. Therefore, since $x \in p^\beta G$, we obtain a commutative diagram

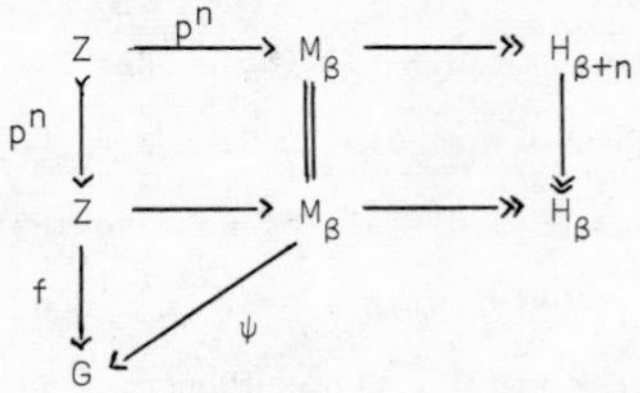

where $f(1) = x$. Hence, $f(p^n) = \psi(p^n) = 0$ which shows that ψ induces a homomorphism ψ^* of $M_\beta/p^nZ = H_{\beta+n}$ into G where $\psi^*(1 + p^nZ) = x$. But $b = 1 + p^nZ$ generates $p^\beta H_{\beta+n}[p^n] = p^\beta H_{\beta+n}$.

<u>Theorem 91</u>. If $e: A \rightarrowtail B \overset{\nu}{\twoheadrightarrow} C$ is p^α-pure, then $p^\beta C[p] = \nu(p^\beta B[p])$ for all $\beta < \alpha$. If C is divisible, then the above condition is also sufficient for e to be p^α-pure.

<u>Proof</u>. Suppose that e is p^α-pure and let $e \in p^\beta C[p]$ where $\beta < \alpha$. By Lemma 90, there is a homomorphism $\phi: H_{\beta+1} \to C$ such that $\phi(b) = c$, where b generates $p^\beta H_{\beta+1}[p]$. Since $\beta+1 \leq \alpha$, then $H_{\beta+1}$ is p^α-projective and so there is a homomorphism $\theta: H_{\beta+1} \to B$ such that $\nu\theta = \phi$. Observing that $\theta(b) \in p^\beta B[p]$ and that $\nu(\theta(b)) = \phi(b) = c$ finishes the first part of the proof.

We proceed in the second part of the proof by induction on α. For $\alpha = 1$, the proof is trivial. If α is a limit ordinal, then $e \in p^\beta \mathrm{Ext}(C,A)$, for each $\beta < \alpha$, implies that $e \in p^\alpha \mathrm{Ext}(C,A)$ since $p^\alpha = \bigcap_{\beta<\alpha} p^\beta$. Hence, suppose that $\alpha = \beta + 1$ and that $\nu(p^\gamma B[p]) = p^\gamma C[p] = C[p]$ (since C is divisible) for all $\gamma \leq \beta$. Therefore, there is a subgroup $S \subseteq p^\beta B[p]$ such that $S \cap A = 0$ and $\nu(S) = C[p]$. We obtain a commutative diagram

$$\begin{array}{ccccc} e: A & \rightarrowtail & B & \xrightarrow{\nu} & C \\ \| & & \downarrow \eta & & \downarrow p \\ e_1: A & \overset{\pi}{\rightarrowtail} & B/S & \xrightarrow{\theta} & C \end{array}$$

where $\pi(a) = a + S$, η is the natural map of B onto B/S and θ is the map induced by the homomorphism $p\nu: B \to C$. (Note that $S \subseteq \operatorname{Ker} p\nu$.) Since $e = pe_1$, it suffices to prove that $\theta(p^{\gamma}(B/S)[p]) = C[p]$ for $\gamma < \beta$. Let $\lambda < \beta$ and $c \in C[p]$. Since C is divisible, we have that $c = pc_1$ and, since $\lambda+1 \leq \beta$, we have that $pc_1 = \nu(s)$ with $s \in S \subseteq p^{\lambda+1}B[p]$. Therefore, $s = pb_1$ where $b_1 \in p^{\lambda}B$. Since $c_1 - \nu(b_1) \in C[p]$, then $c_1 - \nu(b_1) = \nu(s_1)$, $s_1 \in S$. Let $b = b_1 + s_1$. Then $b + S \in p^{\lambda}(B/S)[p]$ and $\theta(b + S) = p\nu(b) = p\nu(b_1) = pc_1 = c$. By our induction hypothesis, it follows that $e_1 \in p^{\beta}\operatorname{Ext}(C,A)$ and thus $e = pe_1 \in p^{\alpha}\operatorname{Ext}(C,A)$.

Definition. A subgroup H of a group G which is maximal with respect to $H \cap p^{\alpha}G = 0$ is called a p^{α}-high subgroup of G. This notion goes back to Irwin [65].

As an application of our previous theorem, we prove...

Theorem 92. If H is a p^{α}-high subgroup of a p-group G, then H is $p^{\alpha+1}$-pure in G. If $\alpha \geq \omega$, then G/H is divisible.

Proof. We may suppose that $p^{\alpha}G \neq 0$; otherwise $H = G$. If $pg \in H$ where $g \notin H$, then the maximality of H with respect to $H \cap p^{\alpha}G = 0$ implies that $mg + h = z \neq 0 \in p^{\alpha}G$. Therefore, m and p are relatively prime and $pz = 0$, since $H \cap p^{\alpha}G = 0$ and $pg \in H$. Hence, $g + uh = uz$ where $um \equiv 1 (\mathrm{mod}\ p^k)$ and $p^k g = 0$. So $pg = p(-uh) \in pH$. Thus, $H \cap pG = pH$. It follows that $(G/H)[p] = \nu(G[p])$, where ν is the natural map of G onto G/H. Since $G[p] = H[p] + p^{\alpha}G[p]$, it follows that, for $\lambda \leq \alpha$, $(G/H)[p] = \nu(G[p]) = \nu(p^{\alpha}G[p]) \subseteq \nu(p^{\lambda}G[p])$. Hence, $\nu(p^{\lambda}G[p]) = (G/H)[p]$ for $\lambda \leq \alpha$. If $\alpha \geq \omega$, then Lemma 8 shows that G/H is divisible and so, by Theorem 91, H is $p^{\alpha+1}$-pure in G. For $\alpha = n < \omega$, we need only prove that $p^m G \cap H = p^m H$ for $m \leq n+1$. This follows from Exercise 10.

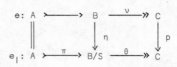

Theorem 93. Let G be a p-group of length α and suppose that $e: A \rightarrowtail B \overset{\nu}{\twoheadrightarrow} G$ is an extension of A by G. If

(i) $G/p^{\beta}G$ is p^{β}-projective for all $\beta < \alpha$,

and (ii) $\nu(p^{\beta}B[p]) = p^{\beta}G[p]$ for all $\beta < \alpha$,

then e is p^{α}-pure.

Proof. The proof is by induction on α. The case $\alpha = 1$ is elementary. First suppose that α is a limit ordinal. By Exercise 5, part (ii) of the hypothesis guarantees that $A \cap p^{\beta}B = p^{\beta}A$ and $\nu(p^{\beta}B) = p^{\beta}G$ for all $\beta < \alpha$. It follows that the sequence $e_{\beta}: A/p^{\beta}A \rightarrowtail B/p^{\beta}B \twoheadrightarrow G/p^{\beta}G$ is exact for all $\beta < \alpha$. Furthermore, it is routine to exhibit (i) and (ii) above for the extensions e_{β}, $\beta < \alpha$. Hence, e_{β} is p^{β}-pure for each $\beta < \alpha$ by our induction hypothesis. By Proposition 87, e is p^{β}-pure for each $\beta < \alpha$ and so e is p^{α}-pure, since $p^{\alpha} = \bigcap_{\beta<\alpha} p^{\beta}$.

If $\alpha = \beta + 1$, then by Lemma 79, G is totally projective. Since $\nu(p^{\beta}B[p]) = p^{\beta}G[p]$, there is a subgroup $S \subseteq p^{\beta}B[p]$ such that $A \cap S = 0$ and $\nu(S) = p^{\beta}G[p]$. Then the induced exact sequence $\varepsilon_1: A \rightarrowtail B/S \overset{\nu*}{\twoheadrightarrow} G/p^{\beta}G$ clearly satisfies (i) for all $\gamma < \beta$. Now let $g + p^{\beta}G \in p^{\gamma}(G/p^{\beta}G)[p] = (p^{\gamma}G/p^{\beta}G)[p]$ for $\gamma < \beta$, that is, $g \in p^{\gamma}G$ with $pg \in p^{\beta}G = p^{\beta}G[p]$. Hence, $\nu(s) = pg$, where $s \in S \subseteq p^{\beta}B[p] \subseteq p^{\gamma+1}B[p]$. Therefore, $s = pb_1$, where $b_1 \in p^{\gamma}B$, and $\nu(b_1) - g \in p^{\gamma}G[p] = \nu(p^{\gamma}B[p])$. Hence, $\nu(b_1) - g = \nu(z)$, $z \in p^{\gamma}B[p]$. Let $b = b_1 - z$. Then $b + S \in p^{\gamma}(B/S)[p]$ and $\nu*(b + S) = \nu(b_1) - \nu(z) + p^{\beta}G = g + p^{\beta}G$. Thus, ε_1 satisfies both (i) and (ii) for all $\gamma < \beta$. But this implies that ε_1 splits since $G/p^{\beta}G$ is p^{β}-projective. Therefore, $B/S = \{A,S\}/S + H/S$ and, hence, $B = A + H$, since $A \cap H \subseteq A \cap S = 0$. Thus, $e = 0 \in p^{\alpha}\mathrm{Ext}(G,A)$.

Lemma 94. Let G be a p-group of length α, where α is a limit ordinal. Then there is a direct sum T of copies of the p-groups H_{β} (generalized Prufer groups), for $\beta < \alpha$, and an epimorphism $\theta: T \twoheadrightarrow G$ such that $\theta(p^{\gamma}T[p]) = p^{\gamma}G[p]$ for all $\gamma < \alpha$.

Proof. Let $g \neq 0 \in p^{\beta}G[p^n]$ and set $H_{(g,n,\beta)} = H_{\beta+n}$. By Lemma 90, there is a homomorphism $\phi_{(g,n,\beta)} : H_{(g,n,\beta)} \to G$ such that $\phi_{(g,n,\beta)} : b \to g$ where b generates $p^{\beta} H_{(g,n,\beta)}[p^{n}]$. Form the direct sum $T = \sum H_{(g,n,\beta)}$ over all triples (g,n,β) and take $\theta: T \to G$ to be the sum of the homomorphisms $\phi_{(g,n,\beta)}$. It is evident that θ is epic and that $\theta(p^{\gamma}T[p]) = p^{\gamma}G[p]$ for all $\gamma < \alpha$.

We now apply Theorem 93 and Lemma 94 to obtain the following important result on totally projectives.

Theorem 95. If G is a totally projective p-group of length α, α a limit ordinal, then G is a direct summand of a direct sum of copies of generalized Prüfer groups H_{α}, $\beta < \alpha$.

Proof. Let $e: K \rightarrowtail T \xrightarrow{\theta} G$ be as in Lemma 94. Then Theorem 93 yields that $e \in p^{\alpha}\mathrm{Ext}(G,K)$. But $p^{\alpha}\mathrm{Ext}(G,K) = 0$, since G is totally projective of length α. Thus, e splits and G is a direct summand of T.

With Theorem 95 established it is now an easy matter to show that the classes of Axiom 3 p-groups and totally projective p-groups coincide.

Theorem 96. Let G be a reduced p-group. Then G is a totally projective p-group if and only if G is an Axiom 3 group.

Proof. If G is an Axiom 3 group, then by Corollary 71, G is in the class $\mathcal{H}$ (= groups which are isomorphic to direct summands of direct sums of copies of generalized Prüfer groups). By Theorem 80, G must be totally projective.

To see that a totally projective p-group G is necessarily an Axiom 3 group, we proceed by induction on the length of G, say length G = α. If α is a nonlimit ordinal, then the induction step is easily accomplished by Theorem 68. If α is a limit ordinal, we simply apply Theorem 95 to finish the proof.

Corollary 97. A p-group G is totally projective if and only if G is a direct summand of a direct sum of copies of generalized Prüfer groups.

Corollary 98. A p-group G is totally projective of length $\leq \Omega$ if and only if G is a reduced d.s.c. p-group (Axiom 2 group).

Corollary 99. Let G be a reduced p-group. Then $p^\beta G$ and $G/p^\beta G$ are totally projective if and only if G is totally projective.

Proof. This is a direct consequence of Theorem 96 and Theorem 74.

We now consider some questions concerning the structure of subgroups of totally projective p-groups; in particular, d.s.c. groups. In the following $\overline{B}$ denotes the torsion completion of $B = \sum_{n<\omega} Z(p^n)$.

Theorem 100. Subgroups of reduced d.s.c. p-groups need not be themselves d.s.c. groups. In particular, $\mathrm{Tor}(H_{\omega+1}, \overline{B})$ is a subgroup of $\sum_{2^{\aleph_0}} H_{\omega+1}$ but is itself not a d.s.c.

Proof. Let $G = \mathrm{Tor}(H_{\omega+1}, \overline{B})$. Then the exact sequence $\overline{B} \rightarrowtail D(\overline{B}) \twoheadrightarrow U$, where $D(\overline{B}) = \sum_{2^{\aleph_0}} Z(p^\infty)$, induces the monomorphism

$$G = \mathrm{Tor}(H_{\omega+1}, \overline{B}) \rightarrowtail \mathrm{Tor}(H_{\omega+1}, \sum_{2^{\aleph_0}} Z(p^\infty)) \simeq \sum_{2^{\aleph_0}} \mathrm{Tor}(H_{\omega+1}, Z(p^\infty)) \simeq \sum_{2^{\aleph_0}} H_{\omega+1}.$$

We also have the exact sequence

$C = \mathrm{Tor}(H_{\omega+1}, B) \rightarrowtail \mathrm{Tor}(H_{\omega+1}, \overline{B}) \twoheadrightarrow \mathrm{Tor}(H_{\omega+1}, \sum_{2^{\aleph_0}} Z(p^\infty))$ which is induced by the pure exact sequence $B \rightarrowtail \overline{B} \twoheadrightarrow \sum_{2^{\aleph_0}} Z(p^\infty)$. Hence, G contains a countable subgroup C such that $|p^\omega(G/C)| = 2^{\aleph_0}$. By Exercise 20, $p^\omega G = \mathrm{Tor}(p^\omega H_{\omega+1}, p^\omega \overline{B}) = \mathrm{Tor}(p^\omega H_{\omega+1}, 0) = 0$. Therefore, if G were a d.s.c., then G would actually be a direct sum of cyclic groups. But if G were a direct sum of cyclic groups, there would be a countable direct summand C_1 of G containing C which yields the contradictory fact $|p^\omega(G/C)| = |p^\omega(C_1/C) + p^\omega(G/C_1)| = |p^\omega(C_1/C) + 0| \leq \aleph_0$.

Our next result takes a step in the positive direction on subgroups of d.s.c.'s. However, we first establish a useful "purification" lemma.

Definition. A subgroup A of the p-group B is called isotype if $A \cap p^\alpha B = p^\alpha A$ for all ordinals α.

Lemma 101. Let G be a p-group and let α be a countable ordinal. If K is an isotype subgroup of G with $K \subseteq C \subseteq G$ and C/K countable, then there is a subgroup H of G containing C such that H/K is countable and $p^\beta G \cap H = p^\beta H$ for all $\beta \leq \alpha$.

Proof. The proof is by induction on α. Suppose first that $\alpha = \beta + 1$.

Observe that $C = \{K, C_o\}$, where C_o is a countable subgroup of G, say $C_o = [x_1, x_2, \ldots, x_n, \ldots]$. If for $i < \omega$, there is some $k_i \in K$ such that $k_i + x_i \in p^{\beta+1}G$, choose a fixed $y_i \in p^{\beta}G$ such that $py_i = k_i + x_i$ with the stipulation that $y_i \in K$ if $x_i \in K$. If there is no such k_i, set $y_i = 0$. Let $L_1 = \{C, y_1, y_2, \ldots\}$. By the induction hypothesis, there is a subgroup $H_1 \supseteq L_1$ such that $|H_1/K| \leq \aleph_0$ and $p^{\gamma}G \cap H_1 = p^{\gamma}H_1$ for all $\gamma \leq \beta$. It follows that $p^{\gamma}G \cap C \subseteq p^{\gamma}H_1$ for $\gamma \leq \beta+1 = \alpha$. Likewise, there is a subgroup H_2 of G containing H_1 such that $|H_2/K| \leq \aleph_0$ and $p^{\gamma}G \cap H_1 \subseteq p^{\gamma}H_2$ for $\gamma \leq \alpha$. Hence, we obtain an ascending sequence of groups $C \subseteq H_1 \subseteq H_2 \subseteq \ldots \subseteq H_n \subseteq \ldots$ such that $|H_n/K| \leq \aleph_0$ and $p^{\gamma}G \cap H_n \subseteq p^{\gamma}H_{n+1}$ for $\gamma \leq \alpha$. Taking $H = \bigcup_{n<\omega} H_n$, we have that $|H/K| \leq \aleph_0$ and $p^{\gamma}G \cap H = p^{\gamma}G \cap (\bigcup_{n<\omega} H_n) = \bigcup_{n<\omega}(p^{\gamma}G \cap H_n) \subseteq \bigcup_{n<\omega} p^{\gamma}H_{n+1} \subseteq p^{\gamma}H$. Thus, for $\gamma \leq \alpha$, $p^{\gamma}G \cap H = p^{\gamma}H$.

Now assume that α is a limit ordinal. The induction hypothesis yields the existence of an ascending chain $C \subseteq H_1 \subseteq \ldots \subseteq H_{\beta} \subseteq \ldots$, $\beta < \alpha$, such that $|H_{\beta}/K| \leq \aleph_0$ and $p^{\gamma}G \cap H_{\beta} = p^{\gamma}H_{\beta}$ if $\gamma \leq \beta < \alpha$. Since α is countable, the chain $[H_{\beta}]_{\beta<\alpha}$ is countable and, therefore, $|H/K| \leq \aleph_0$, where $H = \bigcup_{\beta<\alpha} H_{\beta}$. Clearly, $p^{\gamma}G \cap H = p^{\gamma}H$ for $\gamma \leq \alpha$.

The next lemma is a technical one and is, in fact, the crucial part of the induction step in the theorem that follows.

<u>Lemma 102</u>. Let G be a reduced p-group, $G = \sum_{i \in I} G_i$ where G_i is countable for $i \in I$, and I is an initial segment of the ordinal numbers. Suppose further that G has countable length α and that H is a subgroup of G. Let $J \subseteq I$ and $i_o \in I - J$.

(1) If $(\sum_J G_i) \cap H$ is isotype in H, then there is a subset I_o of I such that $I_o \supseteq [i_o] \cup J$, $|I_o - J| \leq \aleph_0$ and $H \cap \sum_J G_i$ is isotype in H.

(2) If $H/p^{\beta}H$ is a d.s.c., $H/p^{\beta}H = \sum_{i \in I} C_i$ where each C_i is countable, and if $\{H \cap \sum_J G_i, p^{\beta}H\}/p^{\beta}H = \sum_J C_i$, then there is a subset I_1 of I such that $I_1 \supseteq [i_o] \cup J$, $|I_1 - J| \leq \aleph_0$ and $\{H \cap \sum_{I_1} G_i, p^{\beta}H\}/p^{\beta}H = \sum_{I_1} C_i$.

<u>Proof</u>. (1) Let $J_1 = J \cup [i_o]$. We apply Lemma 101 and obtain an isotype subgroup K_1 of H such that $H \cap \sum_{J_1} G_i \subseteq K_1$ and $|K_1/H \cap \sum_J G_i)| \leq \aleph_0$. Clearly, there is a subset J_2 of I such that $|J_2 - J| \leq \aleph_0$ and $K_1 \subseteq H \cap \sum_J G_i$. We again apply Lemma 101 to obtain an isotype subgroup $K_2 \supseteq H \cap \sum_{J_2} G_i$ such that $|K_2/(H \cap \sum_{J_2} G_i)| \leq \aleph_0$. Thus, we continue in this fashion and obtain an increasing sequence $[J_n]_{n<\omega}$ of subsets of I and an increasing sequence $[K_n]_{n<\omega}$ of isotype subgroups of H such that

(i) $|J_n - J| \leq \aleph_0$, $|K_n/H \cap \sum_J G_i| \leq \aleph_0$, $n < \omega$.

(ii) $H \cap \sum_{J_n} G_i \subseteq K_n \subseteq H \cap \sum_{J_{n+1}} G_i$.

Let $I_o = \bigcup_{n<\omega} J_n$ and $K = \bigcup_{n<\omega} K_n$. Then $|I_o - J| \leq \aleph_0$, K is isotype in H, and $K = H \cap \sum_{I_o} G_i$.

The proof of (2) is very similar to the above and so we leave its proof to the reader as an exercise (see Exercise 8).

<u>Proposition 103</u>. Let G be a reduced d.s.c. p-group of length α, α a countable limit ordinal. If H is any subgroup of G such that $H/p^\beta H$ is a d.s.c. for each $\beta < \alpha$, then H is a d.s.c.

<u>Proof</u>. Let I be an initial segment of the ordinal numbers and write $G = \sum_{i \in I} G_i$, G_i countable for each $i \in I$. Also, let $H/p^\beta H = \sum_{i \in I} C_{\beta i}$, $C_{\beta i}$ countable for $i \in I$ and $\beta < \alpha$. There is no loss in generality by using the same index set I since some or all of the $C_{\beta i}$ may be zero. We wish to express I as the monotone increasing union of subsets $[I_\mu]_{\mu \in I}$ such that

(i) $H \cap \sum_{i \in I_\mu} G_i$ is isotype in H.

(ii) $\{H \cap \sum_{i \in I_\mu} G_i, p^\beta H\}/p^\beta H = \sum_{i \in I_\mu} C_{\beta i}$.

(iii) $I_\mu = \bigcup_{\gamma<\mu} I_\gamma$ if μ is a limit ordinal.

(iv) $|I_{\mu+1} - I_\mu| \leq \aleph_0$.

(v) $\mu \in I_{\mu+1}$.

Suppose the collection $[I_\mu]_{\mu<\gamma}$ has been constructed so as to satisfy (i)-(v). If γ is a limit ordinal, we set $I_\gamma = \bigcup_{\mu<\gamma} I_\mu$ and observe that $[I_\mu]_{\mu \leq \gamma}$ satisfies (i)-(v). Suppose that $\gamma = \mu + 1$ and let $J = I_\mu \cup [\mu]$. By Lemma 102(1), there is a subset $J_{o1} \supseteq J$ such that $|J_{o1} - J| \leq \aleph_0$ and $H \cap \sum_{J_{o1}} G_i$ is

isotype in H. We now apply Lemma 102(2) successively to obtain subsets
$J_{o1} \subseteq J_{11} \subseteq \ldots \subseteq J_{\beta 1} \subseteq \ldots$, $\beta < \alpha$, such that $\{H \cap \sum_{J_{\beta 1}} G_i, p^\beta H\}/p^\beta H = \sum_{J_{\beta 1}} C_{\beta i}$.
We define $J_{\alpha 1} = \bigcup_{\beta < \alpha} J_{\beta 1}$ and note that $|J_{\alpha 1} - J| \leq \aleph_o$, since α is countable.
Now start all over again with $J_{o2} \supseteq J_{\alpha 1}$ such that $H \cap \sum_{J_{o2}} G_i$ is isotype in H.
We then obtain $J_{o2} \subseteq J_{12} \subseteq \ldots \subseteq J_{\beta 2} \subseteq \ldots$, $\beta < \alpha$, as before so that
$|J_{\beta 2} - J| \leq \aleph_o$ and $\{H \cap \sum_{J_{\beta 2}} G_i, p^\beta H\}/p^\beta H = \sum_{J_{\beta 2}} C_{\beta i}$. Continuing in this way, we
obtain a chain $J_{o1} \subseteq \ldots \subseteq J_{\beta 1} \subseteq \ldots \subseteq J_{\alpha 1} \subseteq J_{o2} \subseteq \ldots \subseteq J_{\beta 2} \subseteq \ldots \subseteq J_{\alpha 2} \subseteq$
$J_{o3} \subseteq J_{13} \subseteq \ldots \subseteq J_{\beta 3} \subseteq \ldots$ such that

   (a) $J_{\beta n} \subseteq J_{\lambda m}$ if $n < m$ or $n = m$ and $\beta \leq \lambda$.

   (b) $H \cap \sum_{J_{\beta n}} G_i$ is isotype in H.

   (c) $\{H \cap \sum_{J_\beta} G_i, p^\beta H\}/p^\beta H = \sum_{J_\beta} C_i$.

   (d) $|J_{n\lambda} - J| \leq \aleph_o$, $n < \omega$, $\lambda \leq \alpha$.

Let $I_\gamma = I_{\mu+1} = \bigcup_{\beta < \alpha} \bigcup_{n < \omega} J_{\beta n}$. It is routine to check that the collection
$[I_\beta]_{\beta \leq \gamma}$ satisfies (i)-(v). We now have established the existence of the
collection $[I_\mu]_{\mu \in I}$ with properties (i)-(v).

It follows by Proposition 87 that $H \cap \sum_{I_\mu} G_i$ is p^β-pure in H for all
$\beta < \alpha$ and, consequently, p^α-pure, since $p^\alpha = \bigcap_{\beta < \alpha} p^\beta$. However,
$(H \cap \sum_{I_{\mu+1}} G_i)/(H \cap \sum_{I_\mu} G_i)$ is a countable group of length $\leq \alpha$ and is, therefore,
p^α-projective (in fact, the above group is isomorphic to a subgroup of
$\sum_{I_{\mu+1} - I_\mu} G_i$). But then we have a decomposition $H \cap \sum_{I_\mu} G_i = L_\mu + (H \cap \sum_{I_\mu} G_i)$
from which it follows that $H = \sum_{\mu \in I} L_\mu$.

The next theorem was first proved by Hill [53]. Our proof is modeled
after the one presented in [63].

Theorem 104. Let G be a reduced p-group. If G is a d.s.c. and H is an
isotype subgroup of countable length, then H is also a d.s.c.

Proof. Let λ be the length of H. Since $\{H, p^\lambda G\}/p^\lambda G$ is isotype in
$G/p^\lambda G$ (see Exercise 6), we may assume that λ is also the length of G. The
proof is by induction on λ. If $\lambda-1$ exists, then by the induction hypothesis
$H/p^{\lambda-1}H \simeq \{H, p^{\lambda-1}G\}/p^{\lambda-1}G$ is a d.s.c. But $p^{\lambda-1}H$ is a direct sum of cyclic
groups of order p and, therefore, by Corollary 75, H is also a d.s.c.

We may assume that λ is a limit ordinal. By the induction hypothesis, $H/p^{\beta}H$ is a d.s.c. for all $\beta < \lambda$. Thus, by Proposition 103, H is also a d.s.c.

Corollary 105. Let α be a countable ordinal and let G be a reduced d.s.c. p-group. If H is a p^{α}-pure subgroup of length $\leq \alpha$, then H is a d.s.c.

Proof. The conclusion follows from Proposition 86 and Theorem 104.

We now have at hand rather precise structural results for certain classes of reduced p-groups satisfying particular properties and, in addition, have the existence of totally projective p-groups of arbitrary length. For the sake of constructing more general p-groups of arbitrary length and for later application, we present an existence theorem initiated by Megibben [96] and later generalized in [63].

Theorem 106. Let M and A be p-groups such that M has length α and M contains a subgroup H with $H \cap pM = pH$ and with $M[p] = \{H[p], p^{\beta}M\}$ for all $\beta < \alpha$. Further, suppose that $M/H \simeq D(A)/A$. Then there is a p-group G such that $G/p^{\alpha}G \simeq M$, $p^{\alpha}G \simeq A$ and H is a p^{α}-high subgroup of G.

Proof. We take G to be a subdirect sum of M and D with kernels H and A. In other words, we consider G to be a subgroup of $M + D(A)$ such that $\{G,M\} = \{G,D(A)\} = M + D(A)$, $G \cap M = H$ and $G \cap D(A) = A$.

First we observe that $G \cap p(M + D(A)) = pG$. If $x \in M + D(A)$ and $px \in G$, we write $x = g + m$ with $g \in G$ and $m \in M$. Then $pm = px - pg \in pM \cap G = pH$ and we have $px = p(g - h) \in pG$, for some $h \in H$.

Since $M[p] \subseteq \{H[p], p^{\beta}M\}$ for all $\beta < \alpha$, $H[p] \subseteq G[p]$ and D(A) is divisible, $(M + D(A))[p] \subseteq \{G[p], p^{\beta}(M + D(A))\}$ for all $\beta < \alpha$. One may apply Theorem 91 to show that G is p^{α}-pure in $M + D(A)$. Hence, by Proposition 86, $G \cap p^{\beta}(M + D(A)) = p^{\beta}G$ for all $\beta \leq \alpha$. In particular, $p^{\alpha}G = G \cap D(A) = A$. Thus, $G/p^{\alpha}G = G/A = G/G \cap D(A) \simeq \{G,D(A)\}/D(A) \simeq \{M,D(A)\}/D(A) \simeq M$.

Finally, we must show that H is maximal in G with respect to $H \cap A = 0$. It suffices to show that $\{H,g\} \cap A \neq 0$ whenever $g \in G - H$. But if $g \in G - H$, we write $g = m + d$ with $m \in M$ and $d \in D(A)$. Since D(A) is the minimal divisible group containing A, there is a positive integer n such that $0 \neq nd \in A$. Then $nm = ng - nd \in G \cap M = H$ and nd is thus a nonzero element of $\{H,g\} \cap A$.

<u>Lemma 107</u>. If M is any p-group of length $\alpha \geq \omega$, there is a subgroup H of M such that $H \cap pM = pH$, $M[p] \subseteq \{H[p], p^{\beta}M\}$ for all $\beta < \alpha$ and $M/H \simeq Z(p^{\infty})$.

<u>Proof</u>. First suppose that $\alpha = \beta + 1$. Let $z \in p^{\beta}M[p]$, $z \neq 0$, and let $M[p] = S + \{z\}$. We choose H maximal in M with respect to $H[p] = S$. By Exercise 2, $pM \cap H = pH$. Since $z \in p^{\gamma}M$, $\gamma < \alpha$, it follows that $M[p] \subseteq \{H[p], p^{\gamma}M\}$ for all $\gamma < \alpha$. Moreover, $(M/H)[p] = \{z + H\}$, since $pM \cap H = pH$. Hence, $p^{\omega}(M/H)[p] = (M/H)[p]$ and so necessarily $M/H \simeq Z(p^{\infty}) \simeq D(\{z + H\})$.

Let α be a limit ordinal. Since M is not divisible, there is $z \in M[p]$ of finite height. We wish to exhibit a collection of subsocles $[T_{\beta}]_{\beta<\alpha}$ such that

(i) $M[p] = T_{\beta} + p^{\beta}M[p]$, $\beta < \alpha$.

(ii) $T_{\beta} \subseteq T_{\beta'}$ if $\beta \leq \beta'$.

(iii) $z \notin T_{\beta}$ for $\beta < \alpha$.

Suppose that $\gamma < \alpha$ and that the collection $[T_{\beta}]_{\beta<\gamma}$ satisfies (i)-(iii). Write $M[p] = \bigcup_{\beta<\gamma} T_{\beta} + X + p^{\gamma}M[p]$ and $z = t + x + y$ where $t \in {}_{\beta<\gamma}T_{\beta}$, $x \in X$ and $y \in p^{\gamma}M[p]$. Since $z \notin {}_{\beta<\gamma}T_{\beta}$ and since any subgroup of $M[p]$ is a direct summand of $M[p]$, we may arrange X so that $y \neq 0$, that is $z \notin \bigcup_{\beta<\gamma} T_{\beta} + X$. Define $T_{\gamma} = \bigcup_{\beta<\gamma} T_{\beta} + X$ and observe that the collection $[T_{\beta}]_{\beta\leq\gamma}$ satisfies (i)-(iii). By induction, there is a collection $[T_{\beta}]_{\beta<\alpha}$ satisfying (i)-(iii). Then $z \notin \bigcup_{\beta<\alpha} T_{\beta}$ and so we may decompose $M[p] = S + \{z\}$, where $\bigcup_{\beta<\alpha} T_{\beta} \subseteq S$. Clearly, $M[p] \subseteq \{H[p], p^{\beta}M\}$, for $\beta < \alpha$, and as above $H \cap pM = pH$ and $M/H \simeq Z(p^{\infty})$.

<u>Corollary 108</u>. If M is any reduced p-group of length $\alpha \geq \omega$, then there is a group G such that $p^{\alpha}G$ is a nonzero cyclic group and $G/p^{\alpha}G \simeq M$. Moreover, there is a group H such that $p^{\beta}H \simeq M$ where $\beta \geq \omega$.

<u>Proof</u>. The proof is an immediate consequence of Theorem 106 and Lemma 107.

Definition. A subsocle S of a p-group G is called summable if $S = \sum_\alpha S_\alpha$ where $S_\alpha - 0 \subseteq p^\alpha G - p^{\alpha+1} G$, that is, the nonzero elements of S_α have height precisely α. If G[p] is summable, then G is called summable.

The proof of the following result is contained in Exercise 7.

Proposition 109. A countable subsocle of a reduced p-group is summable.

Corollary 110. A reduced countable p-group is a summable group.

Since direct sums of summable p-groups are easily seen to also be summable, we have...

Proposition 111. A reduced d.s.c. p-group is summable.

The following result has very interesting consequences. As always Ω denotes the first uncountable ordinal.

Theorem 112. Let G be a reduced p-group. If $G[p] = S + p^\Omega G[p]$. Then in order for S to be summable, it must be that $p^\Omega G[p] = 0$.

Proof. Suppose that $G[p] = S + p^\Omega G[p]$, where $S = \sum_{\alpha<\Omega} S_\alpha$, $S_\alpha - 0 \subseteq p^\alpha G - p^{\alpha+1} G$, and $p^\Omega G[p] \neq 0$. Let $\pi_\alpha\colon S \to S_\alpha$ be the natural projection associated with the above decomposition of S. Since $p^\Omega G \neq 0$, there is an element $z \in G$ having height exactly Ω. It is easy to inductively define a sequence $y_o, y_1, \ldots, y_i, \ldots$ in G such that, for each $i < \omega$,

(1) $py_i = z$.

(2) y_i has countable height β_i.

(3) $\beta_{i+1} > \beta_i$.

(4) $\pi_\alpha(y_{i+1} - y_i) = 0$ for $\alpha \geq \beta_{i+1}$.

Next let y_ω be an element of G such that $py_\omega = z$ and such that y_ω has countable length $\beta \geq \sup_{i<\omega} \beta_i$. Let $x = y_\omega - y_o$. Then $x \in G[p]$ and there is no loss in generality in assuming that $x \in S$, for we can reselect y_ω. Now it is easy to see that $\pi_\alpha(x) = \pi_\alpha(y_{i+1} - y_i)$ for all $\alpha < \beta_{i+1}$. In particular, $\pi_{\beta_i}(x) \neq 0$ for each i. This contradicts the fact that $\pi_\alpha(x) = 0$ for all but a finite number of α if $x \in S = \sum S_\alpha$ and finishes the proof.

We find in the next theorem an application of McGibben's existence theorem (Theorem 106).

Theorem 113. If G is a summable p-group, then $p^{\Omega}\mathrm{Ext}(Z(p^{\infty}),G) = 0$.

Proof. Suppose that $p^{\Omega}\mathrm{Ext}(Z(p^{\infty}),G) \neq 0$. Then there is a reduced group K containing G as a p^{Ω}-pure subgroup with $K/G \simeq Z(p^{\infty})$. In particular, by Theorem 91, we have that $K[p] \subseteq \{G[p], p^{\alpha}K\}$ for all $\alpha < \Omega$. By Theorem 106, we can find a reduced group H containing G as a p^{Ω}-high subgroup such that $p^{\Omega}H \simeq Z(p)$. Then $H[p] = G[p] + p^{\Omega}H$ and the summability of G implies that of H contrary to the fact that H has length $\Omega+1$ (see Theorem 112).

The above results can be used to exhibit curious properties in the following...

Example. Let G be a p^{Ω}-high subgroup of the totally projective p-group $H_{\Omega+1}$. Since $H_{\Omega+1}/p^{\Omega}H_{\Omega+1} \simeq H_{\Omega}$ and $G \cap p^{\Omega}H_{\Omega+1} = 0$, we may also identify G with a subgroup of the d.s.c. group H_{Ω}. In both cases, $H_{\Omega+1}/G \simeq Z(p^{\infty})$ and $H_{\Omega}/G \simeq Z(p^{\infty})$.

(1) G is not summable.

This is a consequence of the fact that $H_{\Omega+1}[\mu] = G[p] + p^{\Omega}H_{\Omega+1}$ and Theorem 112. Hence, we have that

(2) G is not a d.s.c.

By Theorem 92, G is a $p^{\Omega+1}$-pure subgroup of $H_{\Omega+1}$. One may then apply Theorem 91 and Proposition 86 to the containment $G \subseteq H_{\Omega}$ to obtain

(3) G is p^{Ω}-pure in H_{Ω}; in particular, G is an isotype subgroup subgroup of the d.s.c. H_{Ω} but G itself is not a d.s.c.

Since $G/p^{\alpha}G = G/p^{\alpha}H_{\Omega} \cap G \simeq \{G, p^{\alpha}H_{\Omega}\}/p^{\alpha}H_{\Omega}$ is an isotype subgroup of the d.s.c. $H_{\Omega}/p^{\alpha}H_{\Omega}$, for $\alpha < \Omega$, we may apply Theorem 104 to establish

(4) $G/p^{\alpha}G$ is a d.s.c. for all $\alpha < \Omega$.

It follows that the countability restriction in Theorem 104 cannot be removed.

(5) G is a p^{Ω}-pure subgroup of the p^{Ω}-projective group H_{Ω} but G itself is not p^{Ω}-projective. Thus, p^{Ω} is not a hereditary radical in the sense of Exercise 14.

If G were p^{Ω}-projective, then G would be totally projective since, by (4), $G/p^{\alpha}G$ is a d.s.c. for $\alpha < \Omega$. This would further imply that G is a d.s.c. (contradicting (2)), since G has length = Ω (see Corollary 98). This phenomenon cannot occur for p^{α}-pure subgroups of p^{α}-projectives with $\alpha < \Omega$ (see Exercise 14).

The last property of G that we discuss does exhibit some amount of projectivity for G.

(6) $p^{\Omega}\mathrm{Ext}(G,A) = 0$ if A is a reduced p-group of countable length; in particular, $p^{\Omega}\mathrm{Ext}(G,A) = 0$ for each countable reduced p-group A while $p^{\Omega}\mathrm{Ext}(G,K) \neq 0$ for some reduced p-group K of cardinality $\aleph_1$.

Suppose that e: $A \rightarrowtail M \twoheadrightarrow G$ represents an element of $p^{\Omega}\mathrm{Ext}(G,A)$ and suppose that $p^{\alpha}A = 0$, where $\alpha < \Omega$. By Proposition 86, $A \cap p^{\alpha}M = p^{\alpha}A = 0$ and, by Proposition 87 and Theorem 91, the sequence e*: $A \rightarrowtail M/p^{\alpha}M \twoheadrightarrow G/p^{\alpha}G$ is exact and p^{α}-pure. We remark that Theorem 91 only shows that $p^{\alpha}M[p]$ gets mapped onto $p^{\alpha}G[p]$. However, a simple finite induction argument shows that $p^{\alpha}M$ is mapped onto $p^{\alpha}G$ which yields exactness of e*. But e* splits since $G/p^{\alpha}G$ is p^{α}-projective. It follows that e splits since $A \cap p^{\alpha}M = 0$.

Let e_o: $K \rightarrowtail T \twoheadrightarrow G$ where T is p^{Ω}-projective ($T = \mathrm{Tor}(H_{\Omega},G)$) and e_o is p^{Ω}-pure. Since G is not p^{Ω}-projective, e_o does not split.

The discussion concerning the above example, I believe, clearly demonstrates the complications that arise when one moves from the countable theory in p-groups to the uncountable theory.

Exercises

1. Let G be a reduced p-group. If G is totally projective of length $\leq \Omega$, prove that G is a d.s.c. (Apply Ulm's Theorem!)

2. Let G be a p-group and $S \subseteq G[p]$. If H is maximal in G with respect to $H[p] = S$, then $pG \cap H = pH$.

3. A subsocle S of the p-group G is called <u>dense</u> if $G[p] = \{S, p^nG[p]\}$

for each positive integer n. If S is a dense subsocle of the p-group G and if H is maximal in G with respect to $H[p] = S$, then H is a pure subgroup of G. (Hint: Prove, by induction on n, that $H \cap p^nG = p^nH$ for each positive integer n. Begin with Exercise 2!)

4. If G is a p-group and if $\alpha \leq \omega$, prove that a p^α-high subgroup H of G is pure. (For $\alpha = \omega$, apply Exercise 3.)

5. Let $A \overset{i}{\rightarrowtail} B \overset{\nu}{\twoheadrightarrow} C$ be exact with C a p-group and i the inclusion map. For any ordinal α, prove the equivalence of the following:

(i) $A \cap p^\gamma B = p^\gamma A$ for all $\gamma \leq \alpha$ and $\nu(p^\gamma B) = p^\gamma C$ for all $\gamma < \alpha$.

(ii) $\nu(p^\gamma B[p]) = p^\gamma C[p]$ for all $\gamma < \alpha$. (Use transfinite induction.)

6. Let G be a p-group and let H be an isotype subgroup of G. For any ordinal α, $H/p^\alpha H \simeq \{H, p^\alpha G\}/p^\alpha G$ is an isotype subgroup of $G/p^\alpha G$.

7. Call a subsocle S of a reduced p-group G height finite if $\{h^G(x) : x \in S - 0\}$ is a finite set. Prove that, if the subsocle S is an ascending union $S = \bigcup_{n<\omega} S_n$ where each S_n is height finite, then S is a summable subsocle. Then conclude that a countable subsocle of a reduced p-group G is summable.

8. Prove Lemma 102(2). (Hint: Use Kaplansky's "infinite juggling" technique.)

9. Let G be a reduced p-group and let H be a p^α-high subgroup of G. Use Theorem 91 to show that $H \simeq \{H, p^\alpha G\}/p^\alpha G$ is isomorphic to a p^α-pure subgroup of $G/p^\alpha G$. Moreover, if $\alpha \geq \omega$ and $p^\alpha G \neq 0$, prove that $p^{\alpha+1}\mathrm{Ext}(Z(p^\infty), H) \neq 0$.

10. For $\alpha \leq \omega$, prove that $e: A \rightarrowtail B \twoheadrightarrow C$ is p^α-pure if and only if $A \cap p^\beta B = p^\beta A$ for all $\beta \leq \alpha$. Hence, e is pure exact if and only if $e \in \bigcap_{n<\omega} n\mathrm{Ext}(C,A) \equiv \mathrm{Pext}(C,A)$.

11. As usual, let $e : Z \rightarrowtail M_\alpha \twoheadrightarrow H_\alpha$ represent the functor p^α. Let $T_\alpha = tM_\alpha$. Prove the following:

(a) $Z = p^\alpha M_\alpha$.

(b) T_α is a p-group of length $\leq \alpha$;;in particular, T_α is isomorphic to a subgroup of H_α.

(c) T_α is a p^α-high subgroup of M_α.

(d) The induced monomorphism $T_\alpha \rightarrowtail H_\alpha$ takes T_α onto a p^α-pure subgroup of H_α, for $\alpha \geq \omega$, and $H_\alpha/T_\alpha \approx Z(p^\infty)$. (Use Theorem 91.)

(e) In the p^α-pure resolution in Theorem 84(a), i.e., $K \rightarrowtail F + \mathrm{Tor}(H_\alpha, A) \xrightarrow{\sigma_A} A$, show that $K = N + \mathrm{Tor}(T_\alpha, A)$ where N is free.

(f) If $\alpha < \Omega$, then T_α and K (as in part (e)) are both p^α-projective.

12. Suppose that $K_1 \rightarrowtail P_1 \xrightarrow{\nu_1} A$ and $K_2 \rightarrowtail P_2 \xrightarrow{\nu_2} A$ are both p^α-pure exact and that both P_1 and P_2 are p^α-projective. Prove the following:

(i) There are homomorphisms $g\colon P_1 \to P_2$ and $f\colon P_2 \to P_1$ such that $\nu_1 = \nu_2 g$ and $\nu_2 = \nu_1 f$.

(ii) Define $\theta = 1_{P_1} - fg$ and $\psi = 1_{P_2} - gf$. Then $\theta\colon P_1 \to K_1$ and $\psi\colon P_2 \to K_2$, i.e., show that $\mathrm{Image}\theta \subseteq K_1$ and $\mathrm{Image}\psi \subseteq K_2$.

(iii) Define $\phi\colon P_1 + K_2 \to P_2 + K_1$ by $\phi(p_1, k_2) = (-g(p_1) + k_2, \theta(p_1) + f(k_2))$ and define $\rho\colon P_2 + K_1 \to P_1 + K_2$ by $\rho(p_2, k_1) = (-f(p_2) + k_1, \psi(p_2) + g(k_1))$. Show that $\phi\rho = \rho\phi = $ identity, that is, show ϕ and ρ are both isomorphisms. We conclude that $P_1 + K_2 \approx P_2 + K_1$.

13. If $\alpha < \Omega$ and if K is a p^α-pure subgroup of the p^α-projective group G, then K is p^α-projective. (Hint: Use Exercises 11 and 12.)

14. If p^α-pure subgroups of p^α-projectives are again p^α-projective (as in Exercise 13 with $\alpha < \Omega$), we call p^α hereditary. Let $A \xrightarrow{i} B \xrightarrow{j} C$ be p^α-pure exact. Show that, if p^α is hereditary, then the natural map $i^*\colon p^\alpha\mathrm{Ext}(B,G) \to p^\alpha\mathrm{Ext}(A,G)$ is an epimorphism, and conversely. To accomplish the proof, choose a p^α-projective resolution $K \rightarrowtail P \xrightarrow{\sigma} B$ and let $L = \sigma^{-1}(A)$. Then form the commutative diagram

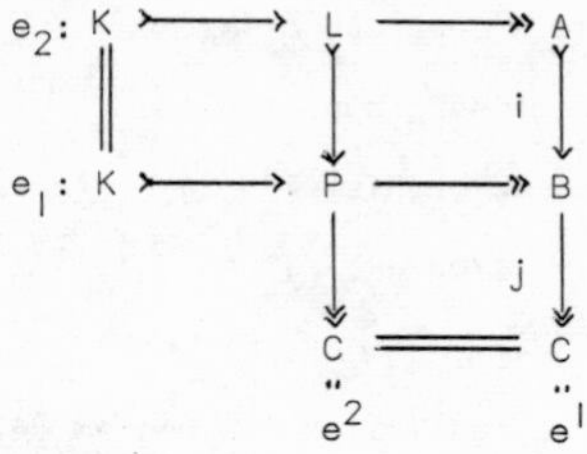

From the p^{α}-purity of e_1 and e^1 derive the p^{α}-purity of e_2 and e^2, respectively. Then consider the induced commutative diagram (via Theorem 89)

$$\begin{array}{ccccc} \mathrm{Hom}(K,G) & \longrightarrow & p^{\alpha}\mathrm{Ext}(B,G) & \longrightarrow & p^{\alpha}\mathrm{Ext}(P,G) = 0 \\ \| & & \downarrow & & \\ \mathrm{Hom}(K,G) & \longrightarrow & p^{\alpha}\mathrm{Ext}(A,G) & \longrightarrow & p^{\alpha}\mathrm{Ext}(L,G). \end{array}$$

15. As in Exercise 14, assume that p^{α} is hereditary and that $A \overset{i}{\rightarrowtail} B \overset{j}{\twoheadrightarrow} C$ is p^{α}-pure exact. Show that the map $j_*\colon p^{\alpha}\mathrm{Ext}(G,B) \to p^{\alpha}\mathrm{Ext}(G,C)$ is an epimorphism. Represent an element of $p^{\alpha}\mathrm{Ext}(G,C)$ by $e\colon C \overset{\pi}{\rightarrowtail} M \twoheadrightarrow G$. We now use Exercise 14 to find a p^{α}-pure extension $e''\colon A \rightarrowtail N \twoheadrightarrow M$ in $p^{\alpha}\mathrm{Ext}(M,A)$ so that $\pi_*(e'') = e'$, i.e., we have a commutative diagram

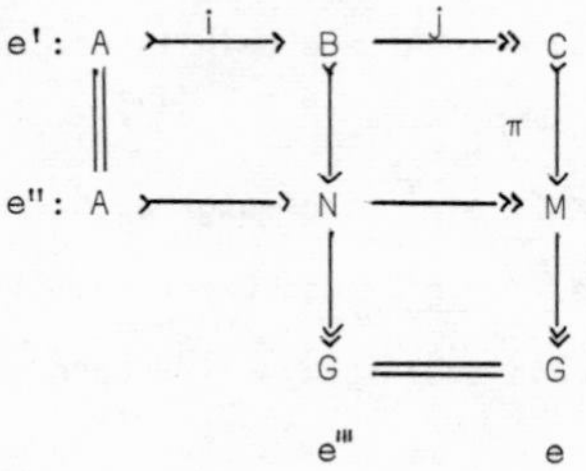

Then prove that $e''' \in p^{\alpha}\mathrm{Ext}(G,B)$ and that $j_*(e''') = e$.

16. If for any p^{α}-pure exact sequence $A \overset{i}{\rightarrowtail} B \overset{j}{\twoheadrightarrow} C$ and any group G we have that the natural maps $i^*\colon p^{\alpha}\mathrm{Ext}(B,G) \to p^{\alpha}\mathrm{Ext}(A,G)$ and $j_*\colon p^{\alpha}\mathrm{Ext}(G,B) \to p^{\alpha}\mathrm{Ext}(G,C)$ are epimorphisms, then $p^{\alpha}\mathrm{Ext}$ is said to be <u>right exact</u>. Prove the equivalence of the following:

(a) Every group G has a p^α-projective resolution $K \rightarrowtail P \twoheadrightarrow G$ with K p^α-projective.

(b) p^α is hereditary.

(c) $p^\alpha\text{Ext}$ is right exact.

17. Show that the following properties are equivalent to those in Exercise 16.

(d) If E is p^α-injective and H is p^α-pure in E, then E/H is p^α-injective.

(e) Every group G has a p^α-pure injective resolution $G \rightarrowtail E \twoheadrightarrow U$ where U is p^α-injective.

18. Show that the p^α-pure exact sequence $T_\alpha \rightarrowtail H_\alpha \twoheadrightarrow Z(p^\infty)$ induces, for any p-group G, a p^α-pure exact sequence $\text{Tor}(T_\alpha,G) \rightarrowtail \text{Tor}(H_\alpha,G) \twoheadrightarrow G$. By iteration show that, for any p-group G, one obtains a long p^α-projective resolution $\ldots \longrightarrow P_n \xrightarrow{d_n} P_{n-1} \xrightarrow{d_{n-1}} \ldots \longrightarrow P_1 \xrightarrow{d_1} P_0 \xrightarrow{\varepsilon} G$ where $P_0 = \text{Tor}(H_\alpha,G)$, in general, $P_n = \text{Tor}(H_\alpha,K_{n-1})$, $K_0 = \text{Tor}(T_\alpha,G)$ and $K_n = \text{Tor}(T_\alpha,K_{n-1})$, $n \geq 1$.

19. If $p^\alpha A = 0$, then $p^\alpha\text{Tor}(A,B) = 0$ for any B. (<u>Hint</u>: See Exercise 12, Chapter IV.)

20. Prove that $p^\alpha\text{Tor}(A,B) = \text{Tor}(p^\alpha A,p^\alpha B)$. (<u>Hint</u>: Construct exact sequence $\text{Tor}(p^\alpha A,p^\alpha B) \rightarrowtail \text{Tor}(A,B) \longrightarrow \text{Tor}(A/p^\alpha A,B) + \text{Tor}(A,B/p^\alpha B)$ and also use generators and relations for Tor(A,B).)

21. If $B \subset p^\alpha G$, then for any group, show the sequence $\ldots\text{Hom}(A,G/B) \xrightarrow{\delta} \text{Ext}(A,B) \longrightarrow p^\alpha\text{Ext}(A,G) \twoheadrightarrow p^\alpha\text{Ext}(A,G/B)$ is exact.

Use the exactness of the sequences $\ldots\text{Hom}(A,G/B) \xrightarrow{\delta} \text{Ext}(A,B) \xrightarrow{j_*} \text{Ext}(A,G) \twoheadrightarrow \text{Ext}(A,G/B)$ and $p^\alpha\text{Ext}(A,G) \rightarrowtail \text{Ext}(A,G) \xrightarrow{\delta_E} \text{Ext}(H_\alpha,\text{Ext}(A,G))$, where $E = \text{Ext}(A,G)$, and use Nunke's Theorem 77 to obtain a commutative diagram

$$\begin{array}{ccccc} & & \mathrm{Ext}(A,B) & & \\ & & \downarrow j_* & & \\ p^{\alpha}\mathrm{Ext}(A,G) & \rightarrowtail & \mathrm{Ext}(A,G) & \xrightarrow{\delta_E} & \mathrm{Ext}(H_{\alpha},\mathrm{Ext}(A,G)) \\ & & \mathrm{Ext}(A,\delta_G) \searrow & & \uparrow\!\uparrow \\ & & & & \mathrm{Ext}(A,\mathrm{Ext}(H_{\alpha},G)) \end{array}$$

Show that $\mathrm{Ext}(A,\delta_G)j_* = 0$ by examining what this map does to an extension $e \in \mathrm{Ext}(A,B)$. Then conclude that $\mathrm{Image}\, j_* \subseteq \mathrm{Ker}\,\delta_E = p^{\alpha}\mathrm{Ext}(A,G)$. This yields sequence $\ldots\mathrm{Hom}(A,G/B) \xrightarrow{\delta} \mathrm{Ext}(A,B) \xrightarrow{j_*} p^{\alpha}\mathrm{Ext}(A,G) \longrightarrow p^{\alpha}\mathrm{Ext}(A,G/B)$ with exactness at $\mathrm{Ext}(A,B)$. To obtain exactness at $p^{\alpha}\mathrm{Ext}(A,G)$ and $p^{\alpha}\mathrm{Ext}(A,G/B)$, show that if $S \rightarrowtail T \twoheadrightarrow U$ is exact with $S \subseteq p^{\alpha}T$, then $S \rightarrowtail p^{\alpha}T \twoheadrightarrow p^{\alpha}U$ is also exact.

22. For any A and G, show that

$$\mathrm{Ext}(A,G)/p^{\alpha}\mathrm{Ext}(A,G) \cong \mathrm{Ext}(A,G/p^{\alpha}G)/p^{\alpha}\mathrm{Ext}(A,G/p^{\alpha}G).$$

23. Prove Theorem 84(b).

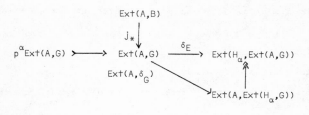

VII. TORSION FREE GROUPS

Our initial task in this chapter is to study torsion free groups of rank one (that is, rank A = 1 if and only if D(A) = Q) and direct sums of such groups. Much of the beginning material here is taken from Baer [8] and also a more complete account can be found in [32].

The first lemma indicates the simplification in the concept of purity resulting from the assumption that a group is torsion free.

Lemma 114. Let G be a torsion free group.

(a) A subgroup H of G is pure if and only if G/H is torsion free.

(b) If H is a subgroup of G, then there is a unique minimal pure subgroup H_* of G which contains H.

(c) If A and B are pure subgroups of G, then $A \cap B$ is also a pure subgroup of G.

Proof. (a) Clear.

(b) Let H_* be the subgroup of G containing H such that $H_*/H = t(G/H)$. Clearly, H_* is the unique minimal pure subgroup of G which contains H.

(c) The exact sequence $A \cap B \rightarrowtail G \longrightarrow G/A + G/B$ shows that $G/(A \cap B)$ is torsion free and thus that $A \cap B$ is pure in G.

Definition. Let G be a torsion free group and $p_1, p_2, \ldots, p_n, \ldots$ the sequence of all rational primes in increasing order of magnitude. For $x \in G$ we define the p_n-height of x by

$$h^G_{p_n}(x) = \begin{cases} m \text{ if } x \in p^m G - p^{m+1} G \text{ (m an integer)} \\ \infty \text{ if } x \in \bigcap_{m<\omega} p^m G \end{cases}$$

The height sequence $H^G(x)$ of x in G is defined by $H^G(x) = (m_1, m_2, \ldots, m_n, \ldots)$ where $m_n = h^G_{p_n}(x)$. We say that two height sequences $(m_1, m_2, \ldots)$ and $(k_1, k_2, \ldots)$ are equivalent if

(i) $m_i = k_i$ whenever $m_i = \infty$ or $k_i = \infty$.

(ii) $m_i = k_i$ for all but a finite number of i.

The relation is easily seen to be an equivalence relation. We further define the <u>type of x in G</u>, denoted by $\tau^G(x)$, to be the equivalence class determined by $H^G(x)$. The set of types admits a partial ordering: $\tau_1 \leq \tau_2$ if and only if there are representative height sequences $(m_1,m_2,\ldots)$ and $(n_1,n_2,\ldots)$ for τ_1 and τ_2, respectively, such that $m_i \leq n_i$ for each i.

<u>Lemma 115</u>. Let G be a torsion free group.

(i) $\tau^G(x) = \tau^G(nx)$ where $n \neq 0$ and $x \in G$.

(ii) If $x,y \in G$ such that $nx = my$, where $n \neq 0$ and $m \neq 0$, then $\tau^G(x) = \tau^G(y)$.

(iii) $\tau^G(x + y) \geq \inf(\tau^G(x),\tau^G(y))$.

(iv) If f is a homomorphism of G into the torsion free group H, then $\tau^G(x) \leq \tau^H(f(x))$.

The proof of this result is left as an exercise (Exercise 1). If A is a rank one torsion free group, it follows from Lemma 115(ii) that each nonzero element in A has the same type τ which we call the <u>type of A</u>. The next result shows that the notion of type is a complete invariant for the class of rank one torsion free groups.

<u>Theorem 116</u>. (Baer) Two rank one torsion free groups are isomorphic if and only if they have the same type.

<u>Proof</u>. The necessity being clear, we assume that A and B are rank one torsion free groups having the same type. Choose nonzero elements $a_o \in A$ and $b_o \in B$. Then a_o and b_o have height sequences in A and B, respectively, which differ in at most a finite number of indices, say $n_1,\ldots,n_s$, and $h^A_{p_{n_i}}(a_o) = e_i$ and $h^B_{p_{n_i}}(b_o) = k_i$ are finite for $i = 1,\ldots,s$. We can solve the equation $p_{n_1}^{e_1} \cdots p_{n_s}^{e_s}x = a_o$ and $p_{n_1}^{k_1} \cdots p_{n_s}^{k_s}y = b_o$ for $a \in A$ and $b \in B$, respectively. Then clearly $\tau^A(a) = \tau^B(b)$ since they are obtained from the above height sequences by replacing e_i and k_i by 0 for $i = 1,\ldots,s$. It follows that an equation $mx = na$, where m and n are non-

zero integers, is solvable in A if and only if my = nb is solvable in B. Since solutions to these equations are unique in a torsion free group, we obtain a one-to-one correspondence $x \to y$ if and only if $mx = na$ and $my = nb$. Clearly, this correspondence yields an isomorphism of A onto B.

Definition. A torsion free group G is called completely decomposable provided that G is a direct sum of rank one groups.

For completely decomposable groups, the above theorem can be improved as follows:

Theorem 117. (Baer) Two completely decomposable torsion free groups are isomorphic if and only if the ranks $r(\tau)$ are equal, where $r(\tau)$ denotes the number of summands of type τ in some decomposition into groups of rank one.

Proof. Since the sufficiency is clear, we suppose that $A = \sum_{i \varepsilon I} A_i$ where each A_i has rank one. Let $A(\tau) = \{x \varepsilon A: \tau^A(x) \geq \tau\}$ and $A^*(\tau) = \{x \varepsilon A: \tau^A(x) > \tau\}$. In this case, $A(\tau)$ and $A^*(\tau)$ are direct sums of those A_i such that A_i has type $\tau_i \geq \tau$ and $\tau_i > \tau$, respectively. Consequently, $A_\tau = A(\tau)/A^*(\tau)$ is the number of summands of type τ and has rank $r(\tau)$. Thus, if B is another completely decomposable group such that rank A_τ = rank B_τ, for each type τ, then clearly $A \approx B$.

We now wish to consider direct summands and more generally pure subgroups of completely decomposable groups. However, two preliminary lemmas are needed.

Lemma 118. Let S be a pure subgroup of a torsion free group G and assume that all the elements of G not in S and all the elements of G/S are of one and the same type τ. Then there is an a in each coset $x + S$ such that $H^G(a) = H^{G/S}(x + S)$.

Proof. Note that $H^G(z) \leq H^{G/S}(x + S)$ for each $z \varepsilon x + S$. The assumption that $\tau^G(z) = \tau^{G/S}(x + S)$ implies that there is a positive integer m such that $H^G(mz) = H^{G/S}(x + S)$, where m contains no prime factor p for which $h_p^G(z) = \infty$. Write $m = p_1^{e_1} \ldots p_r^{e_r}$ and let $u = p_1^{k_1} \ldots p_r^{k_r}$, where

$k_i = h_{p_i}^{G/S}(x + S)$. There exists $y + S \in G/S$ such that $uy + S = x + S$. As above, there exists a positive integer n such that $H^G(nuy) = H^{G/S}(uy + S) = H^{G/S}(x + S)$, where n contains no prime factor p with $h_p^G(uy) = \infty$. By choice of u, we have that $(n,u) = 1$ and so $(n,m) = 1$ (relatively prime). If the integers s,t are such that $ns + mt = 1$, then with $a = (ns)uy + (mt)z = (ns)z + (mt)z + s = z + s$, $s \in S$, we obtain $H^G(a) \leq H^{G/S}(x + S) = \inf(H^G(nuy), H^G(mz)) \leq \inf(H^G(nsuy, H^G(mtz)) \leq H^G(a)$, that is, $H^G(a) = H^G(a + S)$.

Lemma 119. Let S be a pure subgroup of the torsion free group G such that

(1) G/S is of rank one.

(2) G/S is of type τ.

(3) Every element of G not in S is again of type τ.

Then S is a direct summand of G.

Proof. Let $x + S$ be a nonzero element in G/S. By Lemma 118, we may suppose that x was chosen so that $H^G(x) = H^{G/S}(x + S)$. Let $A = \{x\}_*$ (= minimal pure subgroup of G containing $\{x\}$). Then $H^G(x) = H^{G/S}(x + S)$ implies that $G = \{S,A\}$. Since it is clear that $S \cap A = 0$, it follows that $G = S + A$.

Theorem 120. (Baer) Let G be a torsion free group, S a pure subgroup of G such that G/S is of finite rank and assume that all the elements of G not in S are of the same type τ. Then G is the direct sum of S and a completely decomposable group if and only if all the elements of G/T have the same type τ for every pure subgroup T of G with $T \supseteq S$.

Proof. Clearly, there are pure subgroups $[T_i]_{0 \leq i \leq n}$ such that $S = T_0 \subseteq T_1 \subseteq \ldots \subseteq T_n = G$ such that $\operatorname{rank}(T_i/T_{i-1}) = 1$ for $i = 1,\ldots,n$. By Lemma 119, T_{i-1} is a direct summand of T_i for $i \geq 1$. Let $T_i = T_{i-1} + S_i$, $1 \leq i \leq n$. Then $G = S + S_1 + \ldots + S_n$. The proof of the necessity is left as an exercise (Exercise 4).

Definition. A torsion free group G is called homogeneous if for each nonzero x and y in G, we have that $\tau^G(x) = \tau^G(y)$, that is, $\{x\}_* \simeq \{y\}_*$.

Theorem 121. (Baer) A pure subgroup S of a homogeneous, completely decomposable group G is again completely decomposable.

Proof. Let $G = \sum_{\lambda<\Lambda} G_\lambda$, rank $G_\lambda = 1$ for $\lambda < \Lambda$ and define $G_\alpha = \sum_{\lambda<\alpha} G_\lambda$ for each $\alpha < \Lambda$. Let $S_\alpha = S \cap G_\alpha$. Then $S_\alpha \subseteq S_{\alpha+1}$ and either $S_\alpha = S_{\alpha+1}$ or $S_{\alpha+1}/S_\alpha$ is isomorphic to a subgroup of G_α of rank 1. Let τ be the common type of the nonzero rank one pure subgroups of G. If $S_{\alpha+1}/S_\alpha \neq 0$, then we may apply Theorem 120 to see that in any case $S_{\alpha+1} = S_\alpha + B_\alpha$, (note that the purity of S in G guarantees that S is homogeneous of the same type as G) where $B_\alpha = 0$ or rank $B_\alpha = 1$. Hence, $S = \sum_{\alpha<\Lambda} B_\alpha$ and thus S is completely decomposable.

While the above theorem is generally sufficient for our later discussions, we mention the following result (without proof) due to Kulikov [82] and Kaplansky [71] with an improvement in proof due to Fuchs [36].

Theorem 122. A direct summand of a completely decomposable group is completely decomposable.

One should note that Kaplansky's theorem on direct summands of d.s.c. groups (Theorem 65) reduces the problem to considering direct summands of countable completely decomposable groups.

We next consider completely decomposable torsion free groups arising as pure subgroups of torsion free groups. This development has a degree of similarity with that of pure independence and basic subgroups in the theory of p-groups.

Definition. An independent subset S of a torsion free group G is called quasi-pure independent if $\sum_{x\in S} \{x\}_*$ is a pure subgroup of G and $\{x\} = \{x\}_*$ whenever $\{x\}_*$ is cyclic and $x \in S$. Note that $\{S\}_* = \sum_{x\in S}\{x\}_*$ if S is quasi-pure independent. We remark that quasi-pure independence is equivalent to pure independence if G is homogeneous of type $[(0,0,0,\ldots,0,\ldots)]$, that is, $\{x\}_* \simeq Z$ for each $x \neq 0 \in G$. We further remark that nonzero torsion free groups may not contain nonvoid pure independent subsets (even though reduced), e.g., I_p^*. However, it is clear that nonzero torsion free groups contain nonvoid quasi-pure independent subsets. Our first proposition has a standard Zorn's Lemma proof.

Proposition 123. Any quasi-pure independent subset S of a torsion free group G is contained in a maximal quasi-pure independent subset of G.

One might conjecture that the cardinality of a maximal quasi-pure independent subset of a torsion free group is an invariant. Unfortunately, the following example demonstrates that this is not the case.

Example. Let $G = \prod_p I_p^*$ (I_p^* = p-adic group). We exhibit maximal quasi-pure independent subsets S and T of G such that $|S| = 1$ and $|T| \geq \aleph_0$. For each prime p let x_p be an element of G whose p-th coordinate is nonzero and whose other coordinates are all zero. Then $X = [x_p: \text{p a prime}]$ is easily seen to be quasi-pure independent. Hence, let T be a maximal quasi-pure independent subset of G containing X (therefore, $|T| \geq \aleph_0$). From Exercise 12 (Chapter III), $G \simeq \hat{Z}$. So let $\{a\} \simeq Z$ be a pure subgroup of G such that $G/\{a\}$ is divisible and let $S = [a]$. Since G is reduced and $G/\{a\}$ is torsion free and divisible, it is elementary to show that S is a maximal quasi-pure independent subset of G.

Although the above example shows that the cardinality of a maximal quasi-pure independent subset is not a group invariant, we are able to establish a slightly weaker result. The proof of this result is essentially the same as Chase's proof of Theorem 3.1 [17].

Theorem 124. If a torsion free group G contains an uncountable quasi-pure independent subset, then any two maximal quasi-pure independent subsets of G have the same cardinality.

Proof. It suffices to show that, if X and Y are quasi-pure independent subsets of G where $|X| < |Y|$ and $\aleph_0 < |Y|$, then there is a quasi-pure independent subset X_1 containing X such that $|X_1| = |Y|$. Set $H = \sum_{x \in X}\{x\}_*$, $K = \sum_{y \in Y}\{y\}_*$ and $\beta = |Y|$. Then H and K are pure subgroups of G, $|H| < |K|$ and $|K| = \beta$. Let $\overline{G} = G/K$ and $\overline{H} = \{H,K\}/K$. Therefore, $\overline{H} \subseteq \overline{G}$, $\overline{G}$ is torsion free and $D(\overline{G}) = D(\overline{H}) + M$, where M is torsion free and divisible. $D(\overline{G})/\overline{H}$ may be identified with $D(\overline{H})/\overline{H} + M$, in which case

$$t(\overline{G}/\overline{H}) \subseteq t(D(\overline{G})/\overline{H}) = t(D(\overline{H})/\overline{H}).$$

Now $t(D(\overline{H})/\overline{H})$ has cardinality less than β, since β is uncountable and since $|\overline{H}| \leq |H| < \beta$. Observing that $\overline{G}/\overline{H} \simeq G/\{H,K\}$, we have shown that $t(G/\{H,K\})$ has cardinality less than β.

Since β is infinite, we may construct a free group F of rank less than β and an epimorphism ψ: $F \to t(G/\{H,K\})$. Then there is a homomorphism ϕ: $F \to G$ such that ψ is the composition of ϕ with the natural map of G onto $G/\{H,K\}$. Since $|\{H,\phi(F)\}| < \beta$, β is infinite and K is completely decomposable, we may write $K = A + B$ where A and B are completely decomposable, $K \cap \{H,\phi(F)\} \subseteq A$ and $\mathrm{rank}(B) = |B| = \beta$. Observing that $H \cap B \subset (H \cap K) \cap B \subset A \cap B = 0$, set $C = H + B$. Then clearly $\mathrm{rank}(C) = \beta$ and C is completely decomposable. Since B is completely decomposable of cardinality β, B contains a quasi-pure independent subset V of cardinality β. Thus, $X \cup V$ will be a quasi-pure independent subset of G if $C = H + B$ is a pure subgroup of G. Suppose $nx \in C$, $n \neq 0$. Then $nx = h_1 + b_1$ where $h_1 \in H$ and $b_1 \in B$. Therefore, x maps onto an element of finite order in $G/\{H,K\}$. Hence, there is an element $y \in F$ such that $x - \phi(y) \in \{H,K\}$. But then $x - \phi(y) = h_2 + a + b_2$ where $h_2 \in H$, $a \in A$ and $b_2 \in B$. We then have that $h_1 + b_1 = nx = n\phi(y) + nh_2 + na + nb_2$, or that

$$h_1 - n\phi(y) - nh_2 - na = nb_2 - b_1.$$

The left hand side of this equation is easily seen to be in A and the right hand side in B. Thus, both sides are zero and we have that $b_1 = nb_2$. Therefore, $h_1 = nx - nb_2 \in nG \cap H = nH$. It follows that $nx = h_1 + b_1 \in nC$, in which case $x \in C$. Hence, C is a pure subgroup of G and $X \cup V$ is a quasi-pure independent subset of G. Setting $X_1 = X \cup V$, we have that X_1 is a quasi-pure independent subset of G such that $X \subseteq X_1$ and $|X_1| = \beta = |Y|$.

Corollary 125. Let G be a torsion free group and suppose that S and T are infinite maximal quasi-pure independent subsets of G. Then $|S| = |T|$.

Proof. Either $|S| = |T| = \aleph_0$ or we may apply Theorem 124.

It now follows that if a torsion free group G contains maximal quasi-pure independent subsets S and T such that $|S| < |T|$, then any maximal quasi-pure independent subset of G is at most countable; in particular, $|T| \leq \aleph_0$.

We now obtain a relationship between the cardinality of a torsion free group and the cardinality of any maximal quasi-pure independent subset of the group.

Theorem 126. If G is a nonzero torsion free group and if S is a maximal quasi-pure independent subset of G, then $|G| \leq (|S| + 1)^{\aleph_0}$.

Proof. Let $G = G_0 + D$ where G_0 is reduced and D is divisible. Since D is torsion free and divisible, it is elementary to show that the cardinality of any maximal quasi-pure independent subset of D is rank(D). It is also easy to show that $S \cap D$ is a maximal quasi-pure independent subset of D whenever S is a maximal quasi-pure independent subset of G. Hence, it is enough to prove the theorem when D = 0, that is, when G is reduced.

Let E be the n-adic completion of G and let H be the closure of $\{S\}_* = \sum_{x \in S} \{x\}_*$ in the n-adic topology on E. Then H is also a pure subgroup of E and E/H is torsion free and reduced. It follows that H is complete and, hence, that Ext(E/H,H) = 0, that is, E = H + M. Since E is torsion free, E = H + M and G is pure in E, then $H \cap G + M \cap G$ is a pure subgroup of G. Therefore, if $M \cap G \neq 0$, we can choose $y \in M \cap G$ such that $S \cup [y]$ is a quasi-pure independent subset of G. But this contradicts the maximality of S. Therefore, $M \cap G = 0$ and the natural projection π of E onto H is a monomorphism when restricted to G. Hence, $|G| = |\pi(G)| \leq |H|$. Since $\{S\}_*$ is dense in H and since H is Hausdorff, we have by Exercise 16 that $|H| \leq |\{S\}_*|^{\aleph_0} = (|S| + 1)^{\aleph_0}$.

Corollary 127. (Corollary of proof) If S is a maximal quasi-pure independent subset of a reduced torsion free group G, then G is isomorphic to a subgroup of the n-adic completion of $\sum_{x \in S} \{x\}_* = \{S\}_*$.

With the aid of Theorem 126, we can establish a stronger version of Corollary 125 for torsion free groups of cardinality greater than the continuum.

Theorem 128. If G is a torsion free group of cardinality greater than the continuum, then any two maximal quasi-pure independent subsets of G have the same cardinality.

Proof. Theorem 126 implies that any two maximal quasi-pure independent subsets of G must be infinite. Thus, by Corollary 125, any two maximal quasi-pure independent subsets of G have the same cardinality.

We shall interpret some of the above results into information concerning completely decomposable pure subgroups of torsion free groups. Recall from Exercise 21 (Chapter III) that H is a pure-essential subgroup of G provided

(1) H is a pure subgroup of G

and (2) if $C \subseteq G$, $H \cap C = 0$ and $H + C$ pure in G, then $C = 0$.

Theorem 129. Every torsion free group G contains a completely decomposable, pure-essential subgroup C and $|G| \leq |C|^{\aleph_0}$ for any such C.

Proof. Let S be a maximal quasi-pure independent subset of G and let $C = \{S\}_* = \sum_{x \in S} \{x\}_*$. It follows that C is a completely decomposable pure subgroup of G. Moreover, the maximality of S implies that C is pure-essential in G. Indeed, a completely decomposable pure subgroup C of G is pure-essential in G if and only if C contains a maximal quasi-pure independent subset of G. Thus, for any such C we have that $|G| \leq |C|^{\aleph_0}$ by Theorem 126.

As a consequence of the above, we have...

Theorem 130. Let G be a torsion free group. If each pure subgroup of G is indecomposable, then $|G| \leq 2^{\aleph_0}$.

Proof. We may suppose that G is reduced and that C is a completely decomposable pure subgroup of G such that $|G| \leq |C|^{\aleph_0}$. However, rank C must be one (we may assume $G \neq 0$) and, hence, $|C| \leq \aleph_0$. Thus, $|G| \leq (\aleph_0)^{\aleph_0} = 2^{\aleph_0}$.

Corollary 131. Let G be a reduced torsion free group in which each pure subgroup is indecomposable. Then G is isomorphic to a subgroup of $\hat{Z}$.

Proof. Let C be as in the proof of Theorem 130. By Corollary 127, G is isomorphic to a subgroup of $\hat{C}$ and, by Exercise 8, C is isomorphic to a direct summand of $\hat{Z}$.

Corollary 132. If G is an arbitrary group in which each pure subgroup is indecomposable, then $|G| \leq 2^{\aleph_0}$.

Proof. Either G satisfies the hypothesis of Theorem 130 or G is one of $Z(p^n)$ or $Z(p^\infty)$ for some prime p.

Contrary to the above, indecomposable torsion free groups may have "large" cardinality (see Corner [22]). We take only a brief look here at indecomposable groups.

Lemma 133. (a) Every endomorphism of I_p^* (= p-adic group) is given by the left multiplication by a p-adic integer.

(b) If A is a p-pure subgroup of I_p^*, then every endomorphism of A extends to an endomorphism of I_p^*.

Proof. (a) Let ε be an endomorphism of I_p^* and let $\pi = \varepsilon(1)$, where 1 is the identity of the ring I_p^*. Denote by ρ_π the endomorphism of I_p^* defined by $\rho_\pi(x) = \pi x$. Then $\theta = \varepsilon - \rho_\pi$ is an endomorphism of I_p^* such that $\theta(I_p) = 0$. Hence, θ induces a homomorphism $\theta^*\colon I_p^*/I_p \to I_p^*$. However, I_p^*/I_p is divisible which implies that $\theta^* = 0$ and that $\varepsilon = \rho_\pi$.

(b) Since A is p-pure in I_p^*, we have that $t_p(I_p^*/A) = 0$ and, by Exercise 14 (Chapter IV), that $\mathrm{Ext}(I_p^*/A, I_p^*) = 0$. From the exact sequence $A \rightarrowtail I_p^* \twoheadrightarrow I_p^*/A$, we obtain the exact sequence $\mathrm{Hom}(I_p^*, I_p^*) \to \mathrm{Hom}(A, I_p^*) \to \mathrm{Ext}(I_p^*/A, I_p^*) = 0$. It follows that an endomorphism ϕ of A, which can be identified as a homomorphism in $\mathrm{Hom}(A, I_p^*)$, can be extended to an endomorphism of I_p^*.

Theorem 134. (Armstrong [2]) Every p-pure subgroup of I_p^* is indecomposable.

Proof. Suppose that A is a p-pure subgroup of I_p^* and that $A = A_1 + A_2$

where $A_1 \neq 0$. If ϕ is the natural projection of A onto A_1, then ϕ extends to a nonzero endomorphism $\overline{\phi}$ of I_p^* via Lemma 133(b). Therefore, by Lemma 133(a), $\mathrm{Ker}\overline{\phi} = 0$ and thus $A_2 = 0$.

We next consider a method of constructing indecomposable groups of any rank m, $c \leq m \leq 2^c$, where c denotes the cardinality of the continuum. The version presented here is taken from [32].

Lemma 135. (DeGroot [29]) There exists 2^c indecomposable groups G_λ, $\lambda \in \Lambda$, of cardinality c such that $\mathrm{Hom}(G_\lambda, G_\mu) = 0$ for $\lambda \neq \mu$.

Proof. Let $D = Q + \sum_{\gamma \in \Gamma} Q_\gamma$, where $Q_\gamma \simeq Q$, for each $\gamma \in \Gamma$ and $|\Gamma| = c$. Choose $a_\gamma \neq 0 \in Q_\gamma$, for each $\gamma \in \Gamma$, and let $b_\gamma = 1 - a_\gamma$, where, of course, $1 \in Q$. Consider two sets of types of the form

$\mathcal{A} = \{\tau_\gamma : \tau_\gamma$ is of the form $\tau_\gamma = [(0,0,k_3,0,k_5,0,k_7,\ldots)]$ with $k_{2i+1} = 0$ or ∞ (at least one is ∞)$\}$;

$\mathcal{T} = \{\tau_\gamma^* : \tau_\gamma^*$ is of the form $\tau_\gamma^* = [(0,0,0,k_4,0,k_6,0,\ldots)]$ with $k_{2i} = 0$ or ∞ (at least one is ∞)$\}$.

Clearly, $\mathcal{A}$ and $\mathcal{T}$ have cardinality c and can be indexed by Γ. We may select 2^c subsets S_λ, $\lambda \in \Lambda$, of $\mathcal{A}$ such that each of any two different S_λ and S_μ contains a type not belonging to the other. We set up a one-to-one correspondence between $\mathcal{A}$ and $\mathcal{T}$ by putting $\tau_\gamma \longleftrightarrow \tau_\gamma^*$ if $k_{2i-1} = k_{2i}$ for $i = 2,3,\ldots$. Thus, T_λ is defined, for $\lambda \in \Lambda$, as the subset of $\mathcal{T}$ corresponding to S_λ under the above correspondence. We construct G_λ according to S_λ and T_λ as follows:

By Exercise 17, there is a subgroup $A_\gamma \subseteq Q + Q_\gamma$ such that $1, a_\gamma, b_\gamma \in A_\gamma$ of type $\tau = [(0,\infty,0,0,0,\ldots)]$, τ_γ and τ_γ^*, respectively. Let $G_\lambda = \{A_\gamma : \tau_\gamma \in S_\lambda\}$. Then $|G_\lambda| = c$ and it can also be shown that 1, a_γ and b_γ have the same respective type in G_λ as they do in A_γ (by showing that A_γ is pure in G_λ). Moreover, $\{1\}_*$ is the unique pure rank one subgroup of G_λ having type $[(0,\infty,0,0,\ldots)] = \tau$ and, for $\gamma \in \Gamma$, there is no nonzero element of G_λ having type in $G_\lambda \geq \tau_\gamma, \tau_\gamma^*$. These two facts imply that in any decomposition of $G_\lambda = C_1 + C_2$ that, for a fixed γ, a_γ, b_γ, 1 belong to the same C_i. It follows that $C_1 = 0$ or $C_2 = 0$ and thus that G_λ is indecomposable.

Consider $\phi \in \mathrm{Hom}(G_\lambda, G_\mu)$, $\lambda \neq \mu$. By construction, there is a type $\tau_i \in S_\lambda$ not in S_μ and the corresponding $\tau_i^* \in T_\lambda$ not in T_μ. If $a_i \in G_\lambda$ is of type τ_i and $b_i \in G_\lambda$ is of τ_i^*, then $\phi(a_i)$ is of type $\geq \tau_i$ and $\phi(b_i)$ is of type $\geq \tau_i^*$. Since, in general, the types τ_γ, τ_γ^*, τ are incomparable, we must have $\phi(a_i) = 0$ and $\phi(b_i) = 0$. Hence, $\phi(1) = 0$ and so for all other a_j, b_j we have $\phi(a_i) = -\phi(b_i)$. Therefore, the image of a_j in G_μ is of type $\geq \tau_j, \tau_j^*$ which from above shows that only $\phi(a_j) = 0$ is possible. Thus, $\phi = 0$.

Theorem 136. There exist indecomposable groups of any rank $\leq 2^c$.

Proof. Since one may appeal to Theorem 134 for all ranks m, $1 \leq m \leq c$, we only consider m's for which $c \leq m \leq 2^c$. Let $H = \sum G_\lambda$ where the G_λ's are as in Lemma 135 and the sum is taken over m distinct λ's. Define G as the group generated by H and by all elements of the form $\frac{1}{2}(e_\lambda + e_\mu)$ where $e_\lambda \in G_\lambda$, $e_\mu \in G_\mu$ correspond to 1 in the above construction of G_λ. Note that $\frac{1}{2}e_\lambda \in G$.

If $\eta\colon G_\mu \to G$, then $\eta(2G_\mu)$ is a subgroup of H. The projections corresponding to the given direct decomposition of H map $\eta(2G_\mu)$ into G_λ. Because of the choice of the G_λ, these images are zero with the exception of that in G_μ. Thus, $\eta(2G_\mu) \subseteq G_\mu$ and, hence, the image of G_μ under η is a subgroup of G_μ. We conclude that the G_λ are fully invariant subgroups of G.

Assume $G = C_1 + C_2$. Since the G_λ are fully invariant, it follows that $G_\lambda = (C_1 \cap G_\lambda) + (C_2 \cap G_\lambda)$. Since the G_λ are indecomposable, we get that $G_\lambda \subseteq C_1$ or $G_\lambda \subseteq C_2$. If $G_\lambda \subseteq C_1$ and $G_\mu \subseteq C_2$, then consider the element $\frac{1}{2}(e_\lambda + e_\mu) = c_1 + c_2$, $c_i \in C_i$. It follows that $c_1 = \frac{1}{2}e_\lambda$ and $c_2 = \frac{1}{2}e_\mu$ contrary to the fact that $\frac{1}{2}e_\lambda \notin G$. Thus, G is necessarily indecomposable of rank m.

This method of construction yields 2^{2^c} pairwise nonisomorphic indecomposable groups of rank 2^c. In fact, we can construct 2^{2^c} different subsets of cardinality 2^c of the set of all G_λ and for each subset of the G_λ we may apply the above construction.

The remainder of this chapter concerns subgroups and homomorphic images of direct products of copies of Z, in particular, for $\Pi_{\aleph_0} Z$. Our initial result establishes a useful method for deciding whether or not a countable group is free.

<u>Theorem 137</u>. (Pontryagin [32]) A countable torsion free group G is free if and only if each of its subgroups of finite rank is free.

<u>Proof</u>. The necessity being clear, we suppose that G is a countable group with the property that each of its subgroups of finite rank is free. Write $G = [g_1, g_2, \ldots, g_n, \ldots]$ and let $F_n = \{g_1, \ldots, g_n\}_*$. Then $F_1 \subseteq F_2 \subseteq \ldots \subseteq F_n \subseteq \ldots$, $G = \bigcup_{n<\omega} F_n$ and F_n is pure in G for $n < \omega$. Since rank $F_n \leq n$, we have by hypothesis that F_n is free and necessarily finitely generated for $n = 1, 2, \ldots$. Hence, F_{n+1}/F_n is torsion free and finitely generated for each n which implies that F_{n+1}/F_n is free. Therefore, $F_{n+1} = F_n + A_{n+1}$ with A_{n+1} free for $n = 1, 2, \ldots$. So $G = F_1 + A_2 + A_3 + \ldots + A_n + \ldots$ and thus G is free.

<u>Definition</u>. A group G is called <u>$\aleph_1$-free</u> if each countable subgroup of G is free. We remark that, in view of Theorem 137, a group G is $\aleph_1$-free if and only if each subgroup of finite rank is free.

<u>Theorem 138</u>. For any cardinal number α, each subgroup of $\Pi_\alpha Z$ is $\aleph_1$-free.

<u>Proof</u>. In view of the above remarks, it is enough to show that subgroups of $\Pi_\alpha Z$ of finite rank are free. If $A \subseteq \Pi_\alpha Z$ is of rank one, then it is clear that some coordinate projection maps A monomorphically onto a subgroup of Z. If rank $A = n+1$, we may use the same procedure in order to write $A = B + C$ where $C \cong Z$ and rank $B = n$. Thus, an elementary induction completes the proof.

In general, we conclude that subgroups of $\aleph_1$-free are again $\aleph_1$-free. We also note that $\Pi_{\aleph_0} Z$ is an example of an $\aleph_1$-free group that is not free

(Corollary 52). Thus, Pontryagin's Theorem (above) cannot be generalized to larger cardinalities without more restrictive conditions.

Definition. We shall call a group G separable if each pure subgroup of G of finite rank is a free direct summand of G. Therefore, each finitely generated subgroup of G is contained in a free direct summand of G. This definition is more restrictive than the one given by Fuchs [32]. However, it agrees with Fuchs' definition on the class of $\aleph_1$-free groups. We call G $\aleph_1$-separable provided each countable subgroup of G is contained in a free direct summand of G.

Theorem 139. For any cardinal number α, the group $\Pi_\alpha Z$ is separable.

Proof. Let τ be the first ordinal having cardinality α (we may assume that $\alpha \geq \aleph_0$) and let $P = \Pi_{\lambda<\tau}\{e_\lambda\}$, where $\{e_\lambda\} \simeq Z$ for each λ. Let A be a pure rank one subgroup of P. By Theorem 138, $A \simeq Z$, say $A = \{a\}$ where $a = \langle n_\lambda e_\lambda \rangle$. Let n denote the minimum of the absolute values $|n_\lambda|$, $\lambda < \tau$. If $n = 1$, then some $n_\beta = \pm 1$ and $P = A + \Pi_{\lambda\neq\beta}\{e_\lambda\}$. Suppose that $n > 1$ and let $n_\lambda = m_\lambda n + r_\lambda$, where $0 \leq r_\lambda < n$. Let $x_1 = \{m_\lambda e_\lambda\}$ and $x_2 = \{r_\lambda e_\lambda\}$. Then $a = nx_1 + x_2$. Since some $m_\lambda = \pm 1$ and $r_\lambda = 0$, we can write $P = \{x_1\} + P_1$ where P_1 is again a direct product of copies of Z, contains x_2 and the $\min[|r_\lambda| : \lambda < \tau] < n$. An elementary induction argument now completes the proof that A is a direct summand of P. We complete the proof by an induction argument on the rank of A. So assume the result for pure subgroups of rank $\leq n$ and let rank $A = n+1$. Since A is free, by Theorem 138, $A = B + C$ where rank $B = n$ and rank $C = 1$. By our induction hypothesis, $P = B + P_0$. Then $A = B + (A \cap P_0)$ and $\text{rank}(A \cap P_0) = 1$. Therefore, $A \cap P_0$ is a pure rank one subgroup of P which implies that $A \cap P_0$ is a direct summand of P and, hence, also of P_0. Thus, we have that A is a direct summand of P.

A characterization of separable groups is found in...

Theorem 140. A group G is separable if and only if G is isomorphic to a pure subgroup of a direct product of copies of Z.

Proof. First suppose that G is a pure subgroup of a direct product P of copies of Z and suppose that A is a pure subgroup of G of finite rank. Note that A is also pure in P. By Theorem 139, A is a direct summand of P and, hence, also of G. Thus, G is separable.

We now suppose that G is separable. It follows that, given any $x \in G$ such that $H(x) = (0,0,0,\ldots)$, then there is a homomorphism $\theta_x: G \to Z$ such that $\theta_x(x) = 1$. Let $\{e_x\} \simeq Z$ for each x with $H(x) = (0,0,0,\ldots)$ and let $P = \prod_x \{e_x\}$. Define $\theta: G \to P$ by $\theta(g) = \langle\theta_x(g)\rangle$. If $g \neq 0 \in G$, then $g = nx$ for some x with $H(x) = (0,0,\ldots)$ which implies that $\theta(g) \neq 0$. Hence, θ is a monomorphism. Moreover, it is elementary to see that θ preserves height sequences computed with respect to G and P, that is, $\theta(G)$ is pure in P.

An example shows that not all subgroups of direct products of copies of Z need be separable. Let $P = \prod_{\aleph_0} Z$ (= Specker group) and let S denote the corresponding direct sum, that is, S is the subgroup of finitely nonzero sequences in P. Let $G = \{S, 2P\}$. Then G is not separable (see Exercise 18).

Corollary 141. (Corollary to proof) A group G is separable if and only if, for each $x \neq 0 \in G$, there is an $f \in \mathrm{Hom}(G,Z)$ such that f preserves the height sequence of x.

Lemma 142. The Specker group P contains a pure free subgroup F of cardinality $2^{\aleph_0}$ such that P/F is divisible.

Proof. See Exercise 20 (Chapter IV).

Corollary 143. Every maximal pure independent subset of the Specker group contains $2^{\aleph_0}$ elements.

Proof. This result follows from Lemma 142, Theorem 124, and the fact that pure independence is equivalent to quasi-pure independence in P.

Lemma 144. Let C be a countable pure subgroup of the Specker group P that contains the corresponding direct sum $S = \sum_{\aleph_0} Z$ and let U be any countable torsion free group. Then there is a pure subgroup A of P such that $A \supseteq C$ and $A/C \simeq U$.

Proof. Since $S \subseteq C$, C pure in P and P/C is a homomorphic image of P/S, it follows by Lemma 45(1) and Theorem 51 that P/C is a torsion free cotorsion group. Since C is generated by a countable pure independent subset, there is via Corollary 143 a pure free subgroup F of P such that $F = C + B$ where $B \simeq S$. Hence, P/C contains a pure free subgroup B^* isomorphic to S. Then $P/C = H + D$ where $B^* \subseteq H$ and H is Hausdorff and complete in the n-adic topology. It follows that $E(B^*) = E(S)$ is isomorphic to a direct summand of H. Let $U = U_o + V$, where U_o is Hausdorff in the n-adic topology and V is divisible. By Exercise 18 (Chapter IV), U_o is isomorphic to a pure subgroup of $E(S)$ and, hence, also of H. Since the divisible subgroup of P/S is uncountable, it follows that the divisible part D of P/C is also uncountable. Hence, V is isomorphic to a direct summand of D. We may, therefore, identify U with a pure subgroup of P/C; so let $A \supseteq C$ with A/C isomorphic to this pure copy of U in P/C. Finally, note that A is necessarily pure in P.

The lemma that follows is of some independent interest.

Lemma 145. Suppose that a group A of cardinality $\aleph_1$ is an ascending union of countable subgroups $A = \bigcup_{\alpha<\Omega} A_\alpha$ such that $A_\alpha = \bigcup_{\beta<\alpha} A_\beta$ when α is a limit ordinal. If $\lambda < \Omega$ and A is a d.s.c., then A_α is a direct summand of A for some limit ordinal $\alpha > \lambda$.

Proof. We may suppose that ϕ is an isomorphism $\phi: A \rightarrowtail\!\!\!\twoheadrightarrow C = \sum_{\alpha<\Omega} C_\alpha$, where each C_α is countable. We obtain an ascending sequence of countable initial segments I of the ordinals $< \Omega$ such that

(1) $\lambda \in I$.

(2) $\phi(\bigcup_{I_{2n-1}} A_\alpha) \subseteq \sum_{I_{2n}} C_\alpha$, $\phi^{-1}(\sum_{I_{2n}} C_\alpha) \subseteq \bigcup_{I_{2n+1}} A_\alpha$ for $n = 1,2,\ldots$.

(3) $\sup I_n$ is a limit ordinal for each n.

Let $I = \bigcup_{n<\omega} I_n$ and let $\beta = \sup_{n<\omega} I_n$. We note that β is necessarily a limit ordinal $< \Omega$ and that $\phi(\bigcup_{\alpha<\beta} A_\alpha) = \sum_{\alpha<\beta} C_\alpha$. Thus, $A_\beta = \bigcup_{\alpha<\beta} A_\alpha$ is a direct summand of A and $\beta > \lambda$.

We are now in a position to show that the Specker group contains a rather varied collection of pure subgroups. This theorem generalizes one proved by Chase [16]. As usual, Ω denotes the first uncountable ordinal.

Theorem 146. Let $[U_\alpha]_{\alpha<\Omega}$ be a family of countable torsion free groups. Then the Specker group contains a pure subgroup A which is an ascending union of pure subgroups $A = \bigcup_{\alpha<\Omega} A_\alpha$ such that

(1) $A_\alpha \simeq \sum_{\aleph_0} Z$ for each α.

(2) $A_\alpha = \bigcup_{\beta<\alpha} A_\beta$ if β is a limit ordinal.

(3) $A_{\alpha+1}/A_\alpha \simeq U_\alpha$.

Furthermore, if there is a countable ordinal λ such that U_α is not free for each limit ordinal $\alpha > \lambda$, then A is not free.

Proof. Let S denote the finitely nonzero sequences in the Specker group P and let $A_0 = S$. The construction of $A = \bigcup_{\alpha<\Omega} A_\alpha$ by induction on α so that each A_α is pure and (1)-(3) are satisfied is an elementary consequence of Lemma 144. The remainder of the theorem follows directly from Lemma 145 and the fact that countable subgroups of P are free.

The idea for the theorem that follows is taken from a paper of Hill [57] in which he constructs a p-group G of cardinality $\aleph_1$ such that

(a) G is not a d.s.c.,

and (b) each countable subgroup of G is contained in a countable direct summand of G.

Theorem 147. There is an $\aleph_1$-separable group that is not free.

Proof. Let $P = \prod_{\alpha<\Omega}\{e_\alpha\}$ where $\{e_\alpha\} \simeq Z$ for each α. For each limit ordinal $\lambda < \Omega$, let $\sigma_\lambda(n)$ be a sequence of ordinals which increases monotonically to λ and, in addition, has the property that $\sigma_\lambda(n) - 2$ always exists. For λ a limit ordinal, define $c_\lambda^{(n)}$ to be the vector in P with formal sum $c_\lambda^{(n)} = \sum_{i\geq n} (i!/n!) e_{\sigma_\lambda(i)}$. Now let G be the subgroup of P generated by the sets $[e_\alpha]_{\alpha<\Omega}$ and $[c_\lambda^{(n)}]_{n<\omega,\lambda<\Omega}$ (λ a limit ordinal). Further define $G_\beta = G \cap \prod_{\alpha<\beta}\{e_\alpha\}$. Since the vectors in G necessarily vanish for all but a countable number of indices, it follows that $G_\beta = \bigcup_{\alpha<\beta} G_\alpha$ if β if a limit ordinal. Clearly, $G_\beta \subseteq G_{\beta+1}$ for all $\beta < \Omega$.

Let $\beta < \Omega$ and let $\rho_{\beta+1}$ be the natural projection of P onto $\prod_{\alpha<\beta+1}\{e_\alpha\}$

restricted to G. To show that $\rho_{\beta+1}$ is a projection of G onto $G_{\beta+1}$, it suffices to show that $\rho_{\beta+1}(G) \subseteq G_{\beta+1}$. It is clear that $\rho_{\beta+1}([e_\alpha]_{\alpha<\Omega}) \subseteq G_{\beta+1}$ and that $\rho_{\beta+1}(c_\lambda^{(n)}) \subseteq G_{\beta+1}$ for $\lambda \leq \beta+1$. Let $\lambda > \beta+1$. Since $\sigma_\lambda(n) \nearrow \lambda$, it follows that there is an n_o such that $\sigma_\lambda(n) > \beta+1$ for all $n \geq n_o$ which further implies that $\rho_{\beta+1}(c_\lambda^{(m)}) \in \sum_{\alpha<\beta+1}\{e_\alpha\} \subseteq G_{\beta+1}$. Hence, $\rho_{\beta+1}(G) = G_{\beta+1}$ and $G_{\beta+1}$ is a direct summand of G for each $\beta < \Omega$. We conclude that G is $\aleph_1$-separable.

Let α be a limit ordinal $< \Omega$. From the definition of $[c_\alpha^{(n)}]_{n<\omega}$, we obtain the relation $c_\alpha^{(n)} = \sum_{i=1}^{n-1} e_{\sigma_\alpha(i)} + n!c_\alpha^{(n)}$ for $n > 1$. It follows that $c_\alpha^{(1)} \in G_{\alpha+1} - G_\alpha$ and that $c_\alpha^{(1)} + G_\alpha = n!c_\alpha^{(n)} + G_\alpha$ for each n. Hence, the divisible part of $G_{\alpha+1}/G_\alpha$ is nonzero which shows that G_α is not a direct summand of G whenever α is a limit ordinal. Thus, we have that G is not free by Lemma 145.

Our final project in this chapter is to describe the homomorphic images of the Specker group as determined by Nunke [102]. Throughout the remainder of this chapter, P denotes the Specker group and S denotes the finitely nonzero sequences in P. Further, let $G^* = \mathrm{Hom}(G,Z)$ for a group G. We already know that $S^* = P$; our first proposition shows the reverse is also true.

<u>Proposition 148</u>. Let $P = \prod_{n<\omega}\{e_n\}$ and let $f_n \in P^*$, $n < \omega$, be defined by $f_n(<r_m e_m>) = r_m$. Then P^* is free with basis $(f_1, f_2, \ldots)$.

<u>Proof</u>. Let $P_m = \{x = <x_i>: x_i = 0 \text{ for } i < m\}$ for $m = 1,2,\ldots$. Let $f \in P^*$. Then to show that $f \in \{f_1, f_2, \ldots, f_n, \ldots\}$, it suffices to show that $f(P_m) = 0$ for some $m < \omega$. So suppose that $f(P_m) \neq 0$ for infinitely many m. If there is an $n_o < \omega$ such that $f(e_n) = 0$ for all $n \geq n_o$, then f induces a homomorphism $\phi: P_{n_o}/S \cap P_{n_o} \to Z$. But $P_{n_o}/S \cap P_{n_o}$ is algebraically compact, by Theorem 51, and so necessarily $\phi = 0$ which implies that $f(P_{n_o}) = 0$; a contradiction. Hence, $f(e_n) \neq 0$ for infinitely many n. There is no loss in generality in assuming that $f(e_n) \neq 0$ for each $n < \omega$. There is some prime, call it p_1, such that p_1 does not divide $f(e_1)$. Index

the remainder of the primes by $p_2, p_3, \ldots$. Let $b_1 = e_1$. If the integers $b_1, \ldots, b_i$ are already defined, let b_{i+1} be a multiple of $p_1 \ldots p_i e_{i+1}$ such that $f(b_1 + \ldots + b_{i+1})$ is divisible by p_{i+1}. One may choose b_{i+1} in this fashion since p_{i+1} and $p_1 \ldots p_i$ are relatively prime. Then p_1 does not divide $f(e_1)$ but, for $b = \langle b_i \rangle$, $f(b) = f(e_1) + p_1 f(a)$ since p_1 divides $\langle 0, b_2, b_3, \ldots \rangle$. Hence, $f(b) \neq 0$. However, for $n < \omega$, $f(b) = f(b_1 + \ldots + b_i) + p_i f(a^i)$ where $a^i = \langle 0, \ldots, 0, b'_{i+1}, b'_{i+2}, \ldots \rangle$ and $b'_{i+k} = \frac{1}{p_i} b_{i+k}$. Since p_i divides $f(b_1 + \ldots + b_i)$ by the definition of b_i's, p_i divides $f(b)$ for $i < \omega$. This implies that $f(b) = 0$; a contradiction.

If G is a group, there is a natural homomorphism $\sigma_G \colon G \to G^{**} = \mathrm{Hom}(G^*, Z)$ defined by $\sigma_G(g) = g^{**}$ where $g^{**}(f) = f(g)$ for $f \in G^*$. A corollary of the above theorem is...

<u>Corollary 149</u>. The natural maps $\sigma_P \colon P \to P^{**}$ and $\sigma_S \colon S \to S^{**}$ are isomorphisms. For the proof of this result, see Exercise 20. Now suppose that A is a subgroup of P. Then the exact sequence $A \rightarrowtail P \twoheadrightarrow P/A$ induces the exact sequence

$$(1) \qquad (P/A)^* \rightarrowtail P^* \longrightarrow A^*$$

Let A' be the image of $(P/A)^*$ in P^*. Then A' consists of all h in P^* such that $h(A) = 0$, that is, A' is the annihilator of A in P^*. Let B be the image of P^* in A^* so that the sequence

$$(2) \qquad (P/A)^* \rightarrowtail P^* \twoheadrightarrow B$$

is exact. Taking duals again gives a commutative diagram

$$(3) \qquad \begin{array}{ccccc} A & \rightarrowtail & P & \twoheadrightarrow & P/A \\ \lambda \downarrow & & \downarrow \sigma_P & & \downarrow \sigma = \sigma_{P/A} \\ B^* & \rightarrowtail & P^{**} & \longrightarrow & (P/A)^{**} \end{array}$$

with exact rows. The vertical map on the left, denoted by λ, is induced by σ_P and σ. We shall use σ_P to identify P in P^{**}. Then the image of B^* in $P^{**} = P$ is the annihilator A'' of A'. Hence, the image of B^* consists of all $x \in P$ such that $f(x) = 0$ whenever $f \in P^*$ and $f(A) = 0$.

<u>Lemma 150</u>. If $A \subset P$, then $P = A'' + C$. Both A'' and C are direct products of at most countably many copies of Z and $(A''/A)^* = 0$.

<u>Proof</u>. We use the notation in diagrams (1)-(3) above. Since A^* can be embedded in a product of copies of Z and $B \subset A^*$, then B can also be embedded in a product of copies of Z. As an image of P^*, B is countable. These two properties together imply that B is free. Hence, the sequence (2) (above) splits. It follows that the bottom row of (3) also splits. Therefore, $P = A'' + C$ with $A'' \simeq B^*$ and $C \simeq (P/A)^{**}$. Since B and $(P/A)^*$ are free of at most countable rank, their duals B^* and $(P/A)^{**}$, respectively, are products of at most countably many copies of Z.

Suppose that $f: A'' \to Z$ such that $f(A) = 0$. Then f can be extended to P by annihilating C. From the definition of A'', $f(A'') = 0$ which implies that $f = 0$. Hence, $(A''/A)^* = 0$.

Let Z be given the discrete topology and P the associated cartesian product topology. Then Proposition 148 yields that each homomorphism of P into Z is continuous. Therefore, every endomorphism of P is continuous and the product topology on P is independent of the way P is represented as a product of copies of Z. From Lemma 150, if $P = A + C$, then both A and C are products of Z's. Moreover, this splitting is topological and the induced topologies on A and C are the product topologies.

<u>Lemma 151</u>. Let $A \subset P$.

(a) If A has finite rank, then A is closed in the product topology on P induced by the discrete topology on Z.

(b) With the topology as in (a) and $\overline{A}$ denoting the closure of A, if A has infinite rank, there is an isomorphism of $\overline{A}$ with P which carries A onto a subgroup of P containing S.

(c) Again with the topology as in (a), If A is dense in P, there is an automorphism of P which carries A onto a subgroup of P containing S.

<u>Proof</u>. Let $P_n = \{x = \langle x_i \rangle \in P: x_i = 0 \text{ for } i < n\}$. Then $P = P_1, P_2, \ldots$

is a base at zero for the product topology on P induced by the discrete topology on Z. There are elements a^n, $n = 1,2,\ldots$, in A such that

(i) $a_i^n = 0$ for $i < n$.

(ii) $a_n^n = 0$ if and only if $a^n = 0$.

(iii) a_n^n divides x_n for all $x = \langle x_i \rangle \in A \cap P_n$.

In view of (i) and (ii), the nonzero a^n's are independent. If A has finite rank, the set of nonzero a^n's is finite and generates $\overline{A}$ (see Exercise 21). Thus, $A = \overline{A}$ in this case which proves (a).

If A has infinite rank, the set of nonzero a^n is infinite. Let k be a one-to-one correspondence between the positive integers and this set. There is an endomorphism f of P such that $f(x)_i = \sum_n x_n a_i^{k(n)}$. The properties (i)-(iii) show that f is an isomorphism between P and A. If e^n is the element of P whose n-th coordinate is 1 and whose other coordinates are zero, then $f(e^n) = a^{k(n)}$ and is in A. Thus, f^{-1} is the isomorphism required to prove (b).

To prove (c), we observe that, if A is dense in P, it has infinite rank and then we apply (b).

Suppose $A \subset P$. If $f \in P^*$ and $f(A) = 0$, then $f(\overline{A}) = 0$ because f is continuous. We, therefore, have $A \subseteq \overline{A} \subseteq A'' \subset P$. Moreover, in view of Lemmas 150 and 151, $\overline{A}$ and A'' are products and A'' is a direct summand of P.

Lemma 152. Let $A \subset P$ with A a product and $(P/A)^* = 0$. Then the induced map $P^* \to A^*$ is a monomorphism. If $U = A^*/P^*$, then $U^* = 0$ and $P/A \cong \mathrm{Ext}(U,Z)$.

Proof. That $P^* \to A^*$ is a monomorphism is clear. Since A is a product, A^* is free and $\mathrm{Ext}(A^*,Z) = 0$. Dualizing the exact sequence

$$P^* \rightarrowtail A^* \twoheadrightarrow U$$

gives the commutative diagram

$$\begin{array}{ccccccccc}
 & & A & \rightarrowtail & P & \twoheadrightarrow & P/A & & \\
 & & \downarrow \sigma_A & & \downarrow \sigma_P & & \downarrow \tau & & \\
U^* & \rightarrowtail & A^{**} & \longrightarrow & P^{**} & \longrightarrow & \mathrm{Ext}(U,Z) & \longrightarrow & \mathrm{Ext}(A^*,Z) = 0
\end{array}$$

with exact rows. The map τ is induced by σ_A and σ_P. Since A and P are countable products of Z's, both σ_A and σ_P are isomorphisms. Thus, τ is an isomorphism and $U^* = 0$.

The following theorem, which (as mentioned above) is due to Nunke [107], describes the homomorphic images of the Specker group P.

Theorem 153. (Nunke) Each homomorphic image of P is the direct sum of a cotorsion group and the direct product of at most countably many copies of Z.

Proof. Let $A \subset P$. If A has finite rank, let $B = A_*$ (= minimal pure subgroup of P containing A). Then rank B = rank A_* and $P = B + C$ where, by Lemma 150, C is a product of Z's and B is a finitely generated product of Z's. Then $P/A = B/A + C$ and B/A is finite, hence, cotorsion, and C is an at most countable product of copies of Z.

Suppose that A has infinite rank. By Lemma 150, we have that $P = A'' + C$ where $A'' \simeq P$, C is a product of at most a countable number of Z's and $(A''/A)^* = 0$. Then $P/A = (A''/A) + C$. It remains only to describe A''/A.

The inclusions $A \subset \overline{A} \subset A''$ induce an exact sequence

$$\text{(a)} \qquad \overline{A}/A \rightarrowtail A''/A \longrightarrow A''/\overline{A}.$$

Clearly, both $\overline{A}$ and A'' are isomorphic to P. From the exact sequence (a), we obtain an exact sequence

$$\text{(b)} \qquad \mathrm{Ext}(Q,\overline{A}/A) \to \mathrm{Ext}(Q,A''/A) \to \mathrm{Ext}(Q,A''/\overline{A}).$$

Since both A'' and $\overline{A}$ are isomorphic to P and $(A''/\overline{A})^* = 0$, Lemma 152 gives that $A''/\overline{A} \simeq \mathrm{Ext}(U,Z)$ for $U = \overline{A}^*/(A'')^*$. By Corollary 43, $A''/\overline{A}$ is cotorsion; so $\mathrm{Ext}(Q,A''/\overline{A}) = 0$. Since A has infinite rank, there is by Lemma 151, an isomorphism of $\overline{A}$ onto P such that the image of A contains S. Hence, $\overline{A}/A$ is isomorphic to a homomorphic image of P/S. By Lemma 45(1) and Theorem 51, $\overline{A}/A$ is cotorsion which gives that $\mathrm{Ext}(Q,\overline{A}/A) = 0$. Hence, an examination of the exact sequence (b) shows that $\mathrm{Ext}(Q,A''/A) = 0$ which allows by Lemma 40 that A''/A is cotorsion.

<u>Definition</u>. Let $P = \prod_{n<\omega}\{e_n\}$ be the Specker group. J. Los calls a torsion free group G <u>slender</u> if every homomorphism of P into G sends all but a finite number of the e_n into zero.

<u>Theorem 154</u>. (Nunke) A torsion free group is slender if and only if it is reduced, contains no copy of the p-adic integers for any prime p and contains no copy of the Specker group.

<u>Proof</u>. Let I_p^* as usual denote the p-adic completion of Z and define $\phi\colon P \to I_p^*$ so that ϕ takes $x = \langle x_n \rangle$ into the limit of the power series $\sum x_n p^n$ in I_p^*. Observe that ϕ maps e_n onto $p^n \neq 0$ for $n = 1,2,\ldots$. We leave as an exercise that Q is not slender. Since a subgroup of a slender group is slender, the proof of the necessity now easily follows.

Let E be a reduced, torsion free cotorsion group. By Lemma 45(4), E is algebraically compact. If E is nonzero, let A be a pure rank one subgroup of E and H the closure of A in the n-adic topology on E. By Lemma 45(3), H is a direct summand of E. As in Exercise 18 (Chapter IV), one easily shows that $H \simeq A = \prod_{\Lambda} I_p^*$ where Λ contains all primes for which the corresponding component of $\tau = \text{type}(A)$ is finite. Thus, every nonzero reduced, torsion free cotorsion group contains a copy of the p-adic integers for some prime p.

Now suppose that G is a torsion free group that is reduced, contains no copy of the p-adic integers for any prime p and contains no copy of the Specker group. Observe that a group is slender if and only if every homomorphic image of P in it is slender. In view of Theorem 153, the preceding paragraph, and the hypothesis on G, a homomorph of P in G is the product of a finite number of copies of Z. If $f\colon P \to \sum_n Z$, then an elementary adaptation of the proof of Proposition 148 shows that $f(e_n) = 0$ for all but a finite number of n, where $P = \prod_{n<\omega}\{e_n\}$.

<u>Corollary 155</u>. (Sasida [123]) A reduced countable torsion free group is slender.

<u>Corollary 156</u>. An $\aleph_1$-free group is slender if and only if it contains

no copy of the Specker group.

Corollary 157. An $\aleph_1$-separable group is slender.

Proof. Since an $\aleph_1$-separable group is necessarily $\aleph_1$-free, it suffices to show that the Specker group P cannot be a subgroup of an $\aleph_1$-separable group. So suppose that G is $\aleph_1$-separable and that $P \subseteq G$. Then $G = C + K$ where $S \subseteq C \simeq \sum_{\aleph_0} Z$. Let f be the restriction to P of the natural projection of G onto C. Then $f(P) \simeq \sum_{\aleph_0} Z$ since $S \subseteq f(P)$. This situation is impossible by Theorem 153.

Our final remark in this chapter concerns another interesting subgroup of the Specker group P. Let B denote the subgroup of bounded sequences in P, that is, $B = \{x = \langle x_n \rangle \in P: \sup|x_n| < \infty\}$. Specker [128] proved that a subgroup of B of cardinality $\leq \aleph_1$ is necessarily free. Thus, if one assumes the Continuum Hypothesis, then B itself must be free.

Exercises

1. Prove Lemma 115.

2. Let G be a torsion free group and let τ be a given type. Prove that $G_\tau = \{x \in G: \tau^G(x) \geq \tau\}$ is a pure subgroup of G.

3. If τ is a given type, prove there is a rank one torsion free group A such that τ = type of A.

4. Complete the proof of Theorem 120.

5. Let E be a reduced, torsion free cotorsion group (= reduced, torsion free algebraically compact). Prove that S is a maximal quasi-pure independent subset of E if and only if $E/(\sum_{x \in S} \{x\}_*)$ is torsion free and divisible.

6. Prove that the cardinality of a maximal pure independent subset of a separable torsion free group is an invariant of the group.

7. With the aid of Exercise 20 (Chapter IV), prove that every maximal pure independent subset of $P = \Pi_{\aleph_0} Z$ has cardinality $2^{\aleph_0}$.

8. Let A be a reduced rank one torsion free group. Prove that $E(A) \simeq \hat{A}$ is isomorphic to a direct summand of $\hat{Z} = \prod_p I_p^*$.

9. Considering the proof of Lemma 133(a), prove that every endomorphsim of $\hat{Z}$ is given by a left multiplication by an element $x = \langle x_p \rangle \varepsilon \hat{Z}$, where multiplication is componentwise.

10. Considering the proof of Lemma 133(b), prove that every endomorphsim of a pure subgroup A of $\hat{Z}$ extends to an endomorphism of $\hat{Z}$. More generally, show that every homomorphism $\phi: A \to \hat{Z}$ extends to an endomorphism of $\hat{Z}$.

11. Let A be a pure subgroup of $\hat{Z}$ and let $\eta(A)$ be the cardinal number of subgroups of $\hat{Z}$ which are isomorphic to A. Use Exercises 9 and 10, to show that $\eta(A) \leq 2^{\aleph_0}$.

12. Let A be a pure subgroup of $\hat{Z}$ which has a nontrivial decomposition. Show that A contains elements $a,b \ \varepsilon \ \hat{Z}$ such that $a \neq 0 \neq b$ but $a \cdot b = 0$. (<u>Hint</u>: Extend projection of A onto one of its summands to endomorphism of $\hat{Z}$ and remember multiplication in $\hat{Z}$ is componentwise.)

13. Use Exercise 12 to prove: If H is a subgroup of $\hat{Z}$ such that each $x = \langle x_p \rangle \ \varepsilon \ H \subseteq \prod_p I_p^*$ has the property that $x_p \neq 0$ for all primes p, then H_* (= minimal pure subgroup containing H) is purely indecomposable, i.e., each pure subgroup of H is indecomposable.

14. Let $X_p = [x_{p\lambda}]_{\lambda \varepsilon \Lambda}$ be an independent subset of I_p^* of cardinality $2^{\aleph_0}$ for each prime p and let $X = [x_\lambda = \langle x_{p\lambda} \rangle \ \varepsilon \ Z]_{\lambda \varepsilon \Lambda}$. For each subset $M \subseteq X$, let $H_M = \{M\}_*$. Show that $[H_M: M \subseteq X]$ is a collection of 2^c, $c = 2^{\aleph_0}$, set theoretically distinct pure subgroups of Z which are purely indecomposable. Use Exercise 13.

15. Using Corollary 131, Exercise 11, and Exercise 14, show there are exactly 2^c nonisomorphic torsion free, purely indecomposable groups.

16. Let H be a group Hausdorff in the n-adic topology. If A is a subgroup of H such that H/A is divisible, i.e., A is dense in H, prove that $|H| \leq |A|^{\aleph_0}$. (<u>Hint</u>: Map H into $\mathop{X}\limits_{\aleph_0} A$ by associating with h ε H a Cauchy

sequence with entries in A and limit h.)

17. With $k_i = 0$ or ∞ for each i, let the three types τ, τ_1 and τ_1^* be given by $\tau = [(0,\infty,0,0,\ldots)]$, $\tau_1 = [(0,0,k_3,0,k_5,0,\ldots)]$ and $\tau_1^* = [(0,0,0,k_4,0,k_6,0,\ldots)]$ such that $k_{2i-1} = k_{2i}$ for $i = 2,3,\ldots$. Let $Q_i \simeq Q$ for $i = 1,2$ and let $a \neq 0 \in Q_1$ and $b \neq 0 \in Q_2$. Show there is a subgroup $G \subseteq Q_1 + Q_2$ containing a and b so that $\tau^G(a) = \tau_1$, $\tau^G(b) = \tau_1^*$ and $\tau^G(a + b) = \tau$.

18. Let $P = \prod_{\aleph_0} Z$, $S = \sum_{\aleph_0} Z$ and let G be the subgroup of P defined by $G = \{S, 2P\}$. Show that $x = \langle 2,2,2,\ldots \rangle$ generates a pure subgroup of G that is not a direct summand of G. Thus, G is not separable.

19. Carry out the induction step in Theorem 146.

20. Prove Corollary 149.

21. Prove Lemma 151(a) when A has finite rank.

VIII. EXTENSIONS OF GROUPS AND MIXED GROUPS

We now take up the problem of how groups are "put together". Particularly, we will be interested in extensions of torsion groups by torsion free groups and vice versa. Actually though, we are only able to obtain information in rather special cases of the aforementioned problems. This is partly due to the fact that torsion free groups are apparently more complicated (using today's techniques) than torsion groups and so extensions of them or by them seem to be even more complicated.

We begin our discussion by introducing a new concept due to Hill and Megibben [63], that of weak p^α-projectivity. This notion is most important for studying extensions of direct sums of cyclics by direct sums of cyclics and, further, for studying extensions of free groups by torsion groups.

<u>Definition</u>. Let G be a group and let $F_0 \rightarrowtail F \twoheadrightarrow G$ be a free resolution of G. We say that G is a <u>weak</u> <u>p^α-projective</u> if, for each group X, we have that $\delta_X(p^\alpha \mathrm{Hom}(F_0,X)) = 0$, where $\delta_X\colon \mathrm{Hom}(F_0,X) \to \mathrm{Ext}(G,X)$ is the connecting homomorphism. It is easy to check that the above definition is independent of the free resolution used.

Our first lemma is left as an exercise (Exercise 1).

<u>Lemma 158</u>. If F is free, then $\phi \in p^\alpha \mathrm{Hom}(F,H)$ if and only if $\phi(F) \subseteq p^\alpha H$.

<u>Theorem 159</u>. A group G is a weak p^α-projective if and only if for each short exact sequence, $A \rightarrowtail B \twoheadrightarrow G$, for each group H and for each homomorphism $\phi\colon A \to p^\alpha H$, there exists a homomorphism $\psi\colon B \to H$ such that the diagram

$$\begin{array}{ccccc} A & \rightarrowtail & B & \twoheadrightarrow & G \\ {\scriptstyle\phi}\downarrow & & \downarrow{\scriptstyle\psi} & & \\ p^{\alpha}H & \rightarrowtail & H & & \end{array}$$

is commutative.

Proof. Consider the commutative diagram

$$\begin{array}{ccccc} F_o & \rightarrowtail & F & \twoheadrightarrow & G \\ {\scriptstyle\sigma}\downarrow & & \downarrow & & \| \\ A & \rightarrowtail & B & \twoheadrightarrow & G \\ {\scriptstyle\phi_o}\downarrow & & \downarrow & & \\ H & \rightarrowtail & C & \twoheadrightarrow & G \end{array}$$

where the rows are exact, the top row is a free resolution of G and $\phi_o(A) \subseteq p^{\alpha}H$ (ϕ_o is the composition of ϕ with the inclusion of $p^{\alpha}H$ into H). Then $\phi_o\sigma \in p^{\alpha}\mathrm{Hom}(F_o,H)$ by Lemma 158. If G is a weak p^{α}-projective, then the bottom row splits and, hence, ϕ_o extends to a homomorphism $\psi: B \to H$ as desired.

Conversely, assume that G satisfies the condition of the theorem and that $\phi_o \in p^{\alpha}\mathrm{Hom}(F_o,H)$. Then we have a commutative diagram

$$\begin{array}{ccccc} F_o & \rightarrowtail & F & \twoheadrightarrow & G \\ {\scriptstyle\phi_o}\downarrow & & \downarrow & & \| \\ H & \rightarrowtail & C & \twoheadrightarrow & G \end{array}$$

where both rows are exact, the top row is a free resolution of G and $\phi_o(F_o) \subseteq p^{\alpha}H$. But then ϕ_o extends to a homomorphism $\psi: F \to H$ and, therefore, the bottom row splits, that is, $\delta_H(\phi_o) = 0$.

Corollary 160. If $G/p^{\alpha}G$ is a weak p^{α}-projective, then every homomorphism $\phi: p^{\alpha}G \to p^{\alpha}H$ extends to a homomorphism from G to H. In particular, the endomorphisms of $p^{\alpha}G$ extend to endomorphisms of G.

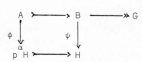

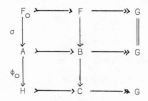

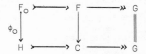

Our next result demonstrates that, in general, weak p^{α}-projectivity may have stronger properties than does p^α-projectivity.

Theorem 161. A subgroup of a weak p^α-projective is itself a weak p^α-projective.

Proof. Suppose that K is a subgroup of the weak p^α-projective group G. Given a free resolution $F_0 \rightarrowtail F \twoheadrightarrow G$, we obtain a commutative diagram

$$\begin{array}{ccccc} F_0 & \rightarrowtail & F_1 & \twoheadrightarrow & K \\ \| & & \downarrow & & \downarrow \\ F_0 & \rightarrowtail & F & \twoheadrightarrow & G \end{array}$$

where the rows are exact and the vertical maps are inclusions. Let $\phi \in p^\alpha \mathrm{Hom}(F_0,H)$. Then $\delta_H(\phi) = 0$, or equivalently, there is a $\phi_1 \in \mathrm{Hom}(F,H)$ such that $\phi_1|F_0 = \phi$. If $\psi = \phi_1|F_1$, then ψ is an extension of ϕ to F_1. Hence, zero is the image of ϕ under the connecting homomorphism $\mathrm{Hom}(F_0,H) \to \mathrm{Hom}(K,H)$.

Since by Exercise 2 a p^α-projective is also a weak p^α-projective, we have...

Corollary 162. If $G/p^\alpha G$ is a subgroup of a d.s.c. of length α, then every endomorphism of $p^\alpha G$ extends to an endomorphism of G.

Theorem 163. If, in the short exact sequence $A \rightarrowtail B \twoheadrightarrow C$, A is weakly p^α-projective and C is weakly p^β-projective, then B is weakly $p^{\alpha+\beta}$-projective.

Proof. Let $A_1 \overset{f_1}{\rightarrowtail} G \twoheadrightarrow B$ be exact and suppose ϕ is a homomorphism of A_1 into $p^{\alpha+\beta}H = p^\beta(p^\alpha H)$. We have a commutative diagram

$$\begin{array}{ccccc} A_1 & \overset{f_2}{\rightarrowtail} & G_1 & \twoheadrightarrow & A \\ \| & & \downarrow \theta & & \downarrow \\ A_1 & \overset{f_1}{\rightarrowtail} & G & \twoheadrightarrow & B \\ & & \downarrow & & \downarrow \\ & & C & = & C \end{array}$$

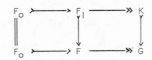

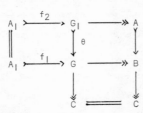

where the rows and columns are exact. Since A is a weak p^α-projective, there is a $\phi_1: G_1 \to p^{\beta}H$ such that $\phi_1 f_2 = \phi$. Since C is a weak p^{β}-projective, there is a $\psi: G \to H$ such that $\psi\theta = \phi_1$. Thus, $\psi f_1 = \psi - f_2 = \phi_1 f_2 = \phi$ and we conclude that B is weakly $p^{\beta+\alpha}$-projective.

Corollary 164. If $F \rightarrowtail G \twoheadrightarrow C$ is exact with F free and C a weak p^{α}-projective, then G is a weak p^{α}-projective.

A characterization of weak p^{α}-projective groups now follows.

Theorem 165. A group G is a weak p^{α}-projective group if and only if G is the extension of a free group by a subgroup of a p^{α}-projective p-group.

Proof. The sufficiency follows immediately from Exercise 2 and Corollary 164. Let us assume that G is a weak p^{α}-projective and let $F_0 \rightarrowtail F \twoheadrightarrow G$ be a free resolution of G. Let $Z \rightarrowtail M_{\alpha} \twoheadrightarrow H_{\alpha}$ be the representative sequence for p^{α} and form the direct sum $\sum_m Z \rightarrowtail \sum_m M_{\alpha} \twoheadrightarrow \sum_m H_{\alpha}$, where $F_0 \simeq \sum_m Z$. Since $\sum_m Z = p^{\alpha}(\sum_m M_{\alpha})$ and since G is a weak p^{α}-projective, there is by Theorem 159 a commutative diagram

$$\begin{array}{ccccc} F_0 & \rightarrowtail & F & \twoheadrightarrow & G \\ \downarrow & & \downarrow \psi & & \downarrow \theta \\ \sum_m Z & \rightarrowtail & \sum_m M_{\alpha} & \twoheadrightarrow & \sum_m H_{\alpha} \end{array}$$

It is easy to check that $\mathrm{Ker}\theta \simeq \mathrm{Ker}\psi$ is free and that $\mathrm{Image}\theta$ is a subgroup of the p^{α}-projective group $\sum_m H_{\alpha}$.

Corollary 166. A torsion group is a weak p^{α}-projective group if and only if it is a subgroup of a p^{α}-projective group.

Hill and Megibben [63] used the above results to give the characterizations below for extensions of direct sums of cyclic groups by direct sums of cyclic groups.

Theorem 167. If G is a p-primary group such that $p^{\alpha}G$ is a subgroup of a reduced d.s.c. p-group and $G/p^{\alpha}G$ is a subgroup of a d.s.c. p-group of countable length, then G is a subgroup of a reduced d.s.c. p-group.

Proof. The proof of this result follows from Theorem 163 and

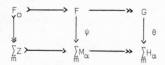

Corollary 166.

Let u_1 denote the radical $u_1G = G' = \bigcap_{n<\omega} nG = \bigcap_p p^\omega G$, that is, $u_1 = \bigcap_p p^\omega$. One easily sees from the proof of Theorem 69 that u_1 has a representative sequence $e: Z \rightarrowtail M \twoheadrightarrow H = \sum_p H_{\omega,p}$ where $H_{\omega,p}$ denotes the generalized Prufer p-group $H_\omega = \sum_{n<\omega} Z(p^n)$. Actually, $e = \langle e_{\omega,p} \rangle_p \in \prod_p \mathrm{Ext}(H_{\omega,p}, Z)$, where $e_{\omega,p}: Z \rightarrowtail M_{\omega,p} \twoheadrightarrow H_{\omega,p}$ represents p^ω. By Exercise 3, the u_1-projective groups are just the pure projectives, that is, direct sums of cyclic groups. Furthermore, it is an easy matter to check that results 158, 159, 161, and 165 hold for the cotorsion function u_1. Finally, note that $u_1M = M' = Z$.

<u>Theorem 168</u>. (Hill-Megibben) A group G is the extension of a direct sum of cyclic groups by a direct sum of cyclic groups if and only if G is the subgroup of a group K such that both K' and K/K' are direct sums of cyclic groups.

<u>Proof</u>. If $G \subseteq K$ and both K' and K/K' are direct sums of cyclic groups, then we have the short exact sequence $G \cap K' \rightarrowtail G \twoheadrightarrow G/G \cap K' \simeq \{G,K'\}/K'$ where both $G \cap K'$ and $\{G,K'\}/K'$, by Theorem 24, are direct sums of cyclic groups. On the other hand, suppose $A \rightarrowtail G \xrightarrow{\nu} B$ is exact, where A and B are direct sums of cyclic groups. Let $e: Z \rightarrowtail M \twoheadrightarrow H$ be the representing sequence (see above) for u_1. We take L to be the direct sum of an appropriate number of copies of M and generalized Prufer groups $H_{\omega 2}$ ($\omega 2 = \omega + \omega$) for the relevant primes so that there is a monomorphism ρ of A into M'. (Note that $p^\omega H_{\omega 2} = H'_{\omega 2} = u_1 H_{\omega 2}$ where p is the relevant prime for $H_{\omega 2}$.) Observe that M/M' is a direct sum of finite cyclic groups. Since B is a weak u_1-projective, we obtain a commutative diagram

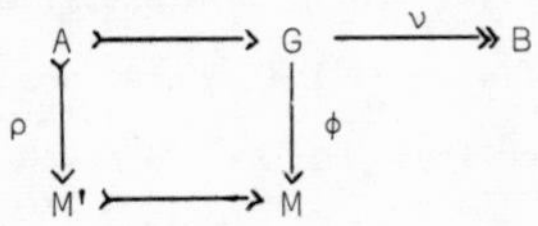

Let K = M + B and define $\theta: G \to M + B$ by: $\theta(g) = (\phi(g), \nu(g)) \in M + B$. It is clear that θ is a monomorphism and that K' = M' and $K/K' \simeq M/M' + B$

are both direct sums of cyclic groups.

Corollary 169. A p-primary group G is the extension of a direct sum of cyclic groups by a direct sum of cyclic groups if and only if G is a subgroup of a d.s.c. p-group of length at most $\omega 2$.

Before continuing our discussion of extensions, we need to establish a theorem which is crucial to several later results.

Definition. A group G is said to have finite torsion free rank provided G/tG has finite rank.

Theorem 170. If $[M_i]_{i\varepsilon I}$ is a family of groups of finite torsion free rank such that the torsion free subgroups of each M_i are free, then the torsion free subgroups of $M = \sum_{i\varepsilon I} M_i$ are free.

Proof. The proof is divided into three cases, namely, when I is finite, countable and finally when I is uncountable. We may, of course, assume that I is an initial segment of the ordinal numbers.

Case I. I is finite. This case proceeds by induction on $|I|$. For $|I| = 1$, the result follows by hypothesis. So suppose the result holds for $|I| \leq n$ and consider $M = \sum_{i=1}^{n} M_i$. Let A be a torsion free subgroup of M and let θ be the natural projection of M onto M/tM where θ restricted to M_i is the natural projection of M_i onto M_i/tM_i. Also, let $B = \{a \varepsilon A: \theta(a) \varepsilon \sum_{i=1}^{n} \theta(M_i)\}$ and let π be the natural projection of M onto $\sum_{i=1}^{n} M_i$. Suppose that $b \varepsilon B \cap \mathrm{Ker}\pi = B \cap M_{n+1}$. Since $\theta(b) \varepsilon \sum_{i=1}^{n} \theta(M_i)$, it follows that $b \varepsilon tM_{n+1} \subseteq tM$. Therefore, $b = 0$ and $B \simeq \pi(B) \subseteq \sum_{i=1}^{n} M_i$. Hence, B is free by the induction hypothesis. Therefore, $B = \sum_{i=1}^{n} \{b_j + y_j\}$ where $b_j \varepsilon \sum_{i=1}^{n} M_i$ and $y_j \varepsilon M_{n+1}$. Since $\theta(b_j + y_j) \varepsilon \sum_{i=1}^{n} \theta(M_i)$, it follows that $y_j \varepsilon tM_{n+1}$ for $j = 1,\ldots,k$. Let m be a positive integer such that $my_j = 0$ for $j = 1,\ldots,k$ and let ρ be the natural projection of M onto M_{n+1}. Set $\phi = m\rho$. It is easily verified that $A \cap \mathrm{Ker}\phi = B$ and that $\phi(A)$ is a torsion free subgroup of M_{n+1}. Hence, $\phi(A)$ is free by hypothesis and thus $A \simeq B + \phi(A)$ is free.

Case II. I is countable. Let A be a countable subgroup of $M = \sum_{i=1}^{n} M_i$.

Since M_i/tM_i has finite rank for each $i \in I$, A is necessarily countable. Therefore, by Pontryagin's Theorem (137), it is enough to prove that A is free when the rank of A is finite. However, in this case, A is isomorphic to a torsion free subgroup of $\sum_J M_i$, where J is a finite subset of I. Thus, by Case I, A is free.

Case III. I is uncountable. We may assume that $o \in I$ and that $M_o = 0$. Let A be a torsion free subgroup of M. Since $|M/tM| \le |I|$, it follows that $|A| \le |I|$. Thus, we may label the elements of A with ordinals in I starting with $a_o = 0$ (we label $o \in A$ repeatedly if necessary). Again let θ be the natural map of M onto M/tM so that, restricted to M_i, θ is the natural map of M_i onto M_i/tM_i, for $i \in I$. We wish to express A and a subset of I as unions of well-ordered monotone sequences $[A_\alpha]_{\alpha \varepsilon I}$ and $[I_\alpha]_{\alpha \varepsilon I}$, respectively, such that $A_o = \{a_o\} = 0$ and $I_o = [0]$ and such that

(i) $|I_{\alpha+1} - I_\alpha| \le \aleph_0$.

(ii) $I_\alpha = \bigcup_{\gamma<\alpha} I_\gamma$ and $A_\alpha = \bigcup_{\gamma<\alpha} A_\gamma$, if α is a limit ordinal.

(iii) $a_\alpha \in A_{\alpha+1}$.

(iv) $A_\alpha = A \cap (\sum_{I_\alpha} M_i) = \{a \in A: \theta(a) \in \sum_{I_\alpha} \theta(M_i)\}$.

Suppose that the I_α's and the A_α's satisfying (i)-(iv) have been chosen for all $\alpha < \beta$, $\beta \in I$. If β is a limit ordinal, then we need only set $I_\beta = \bigcup_{\alpha<\beta} I_\alpha$ and $A_\beta = \bigcup_{\alpha<\beta} A_\alpha$. We may assume that $\beta-1$ exists. Let $B_1 = \{A_{\beta-1}, a_{\beta-1}\}$ and let L_1 be the smallest subset of I such that $B_1 \subseteq \sum_{L_1} M_i$. Clearly, $I_{\beta-1} \subseteq L_1$ and $|L_1 - I_{\beta-1}| \le \aleph_0$. In general for $n > 1$, we let L_n be the smallest subset of I such that $B_n = \{a \in A: \theta(a) \in \sum_{L_{n-1}} \theta(M_i)\} \subseteq \sum_{L_n} M_i$. It is elementary that $B_n \subseteq B_{n+1}$ and $L_n \subseteq L_{n+1}$ for each n. Set $I_\beta = \bigcup_{n<\omega} L_n$ and set $A_\beta = \bigcup_{n<\omega} B_n$. Since (ii) and (iii) are clearly true for $[I_\alpha]_{\alpha\le\beta}$ and $[A_\alpha]_{\alpha\le\beta}$, we need only verify (i) and (iv). We also may assume that $a_{\beta-1} \notin A_{\beta-1}$ so that $L_1 \ne I_{\beta-1}$. To show that $|I_\beta - I_{\beta-1}| \le \aleph_0$, it suffices to show that $|L_n - I_{\beta-1}| \le \aleph_0$ for each n. Since we have already observed that $|L_1 - I_{\beta-1}| \le \aleph_0$, we suppose that $|L_n - I_\beta| \le \aleph_0$ and consider L_{n+1}. By definition of L_{n+1}, it is enough to show that $|B_{n+1}/A_{\beta-1}| \le \aleph_0$. Let

π be the natural projection of M onto $\sum_{L_n - I_{\beta-1}} M_i$. Clearly, $A_{\beta-1} \subseteq \mathrm{Ker}\pi \cap B_{n+1}$. Suppose that $x \in B_{n+1}$ and that $mx \in \mathrm{Ker}\pi \cap B_{n+1}$, where m is a nonzero integer. Then $x = y + \omega$ where $y \in \sum_{I_{\beta-1}} M_i$ and $w \in \sum_{L_{n+1} - I_{\beta-1}} M_i$. Since $\pi(\omega) \in t(\sum_{L_n - I_{\beta-1}} M_i)$ and since $\theta(x) \in \sum_{L_n} \theta(M_i)$, we have that $\omega \in t(\sum_{L_{n+1} - I_{\beta-1}} M_i)$ which implies that $\theta(x) = \theta(y) \in \sum_{I_{\beta-1}} \theta(M_i)$. By (iv), $x \in A_{\beta-1}$. Hence, $\mathrm{Ker}\pi \cap B_{n+1} = A_{\beta-1}$ and $\pi(B_{n+1})$ is torsion free. Since $|L_n - I_{\beta-1}| \leq \aleph_0$, it follows from the definition of the M_i's that $|\pi(B_{n+1})| \leq \aleph_0$. Thus, $|B_{n+1}/A_{\beta-1}| \leq \aleph_0$ and, hence, $|L_{n+1} - I_{\beta-1}| \leq \aleph_0$. Thus, (i) holds for $\alpha \leq \beta$. Now if $x \in A \cap (\sum_{I_\beta} M_i)$, then $x \in A \cap (\sum_{L_n} M_i)$ for some n. Therefore, $\theta(x) \in \sum_{L_n} \theta(M_i)$ which implies that $x \in B_{n+1} \subseteq A_\beta$. Since $A_\beta \subseteq A \cap (\sum_{I_\beta} M_i)$, we have that $A_\beta = A \cap (\sum_{I_\beta} M_i)$. We also have that $A_\beta = A \cap (\sum_{I_\beta} M_i) \subseteq \{a \in A: \theta(a) \in \sum_{I_\beta} \theta(M_i)\}$. If $\theta(a) \in \sum_{I_\beta} \theta(M_i)$ where $a \in A$, then $\theta(a) \in \sum_{L_n} \theta(M_i)$ for some n. By definition, $a \in B_{n+1} \subseteq A_{\beta+1}$. Hence, $A_\beta = A \cap (\sum_{I_\beta} M_i) = \{a \in A: \theta(a) \in \sum_{I_\beta} \theta(M_i)\}$. Thus, $[A_\alpha]_{\alpha \leq \beta}$ and $[I_\alpha]_{\alpha \leq \beta}$ satisfy (i)-(iv) and the induction is completed.

We now establish

(v) A_α is a direct summand of $A_{\alpha+1}$ and $A_{\alpha+1} = A_\alpha + F_\alpha$ where F_α is free.

Let π_α be the natural projection of M onto $\sum_{I_{\alpha+1} - I_\alpha} M_i$ (we may assume that $I_{\alpha+1} \neq I_\alpha$). Suppose that $x \in A_{\alpha+1}$ such that $\pi(x) \in t(\sum_{I_{\alpha+1} - I_\alpha} M_i)$. Then $x = y + \omega$ where $y \in \sum_{I_\alpha} M_i$ and where $\omega \in t(\sum_{I_{\alpha+1} - I_\alpha} M_i)$. This implies that $\theta(x) = \theta(y) \in \sum_{I_\alpha} \theta(M_i)$ which implies by (iv) that $x \in A_\alpha$. Hence, $A_{\alpha+1} \cap \mathrm{Ker}\pi_\alpha = A_{\alpha+1} \cap (\sum_{I_\alpha} M_i) = A \cap (\sum_{I_\alpha} M_i) = A_\alpha$ and $\pi_\alpha(A_{\alpha+1})$ is a torsion free subgroup of $\sum_{I_{\alpha+1} - I_\alpha} M_i$. From Case II of the proof, $\pi(A_{\alpha+1})$ is free. This establishes (v). Since A_o is free, then (ii) and (v) imply that A is free.

Numerous consequences of the above result now follow.

Lemma 171. Let p be a prime, α an ordinal number and $e_\alpha: Z \rightarrowtail M_\alpha \twoheadrightarrow H_\alpha$ the sequence representing p^α. Then each torsion free subgroup of $\sum_m M_\alpha$ is free for any cardinal number m.

Proof. From Theorem 170, it suffices to show that each torsion free subgroup of M_α is isomorphic to a subgroup of Z (note that this implies M_α has torsion free rank one). Hence, let A be a torsion free subgroup of M_α. We may assume that $A \neq 0$. Then $A/A \cap Z$ is isomorphic to a subgroup of H , i.e., $A/A \cap Z$ is a reduced p-group. Therefore, $D(A) = D(A \cap Z) \simeq D(Z) = Q$. Hence, A has rank one and $A/A \cap Z$ isomorphic to a reduced subgroup of $(Q/Z)_p = Z(p^\infty)$. Thus, $A/A \cap Z$ is a cyclic p-group from which it follows that $A \simeq Z$.

We now prove our main result concerning extensions of free groups by torsion groups.

Theorem 172. Let p be a fixed prime and α an arbitrary ordinal. If a torsion free group G contains a free subgroup F with G/F a weak p^α-projective p-group, then G is free.

Proof. Let e_α: $Z \rightarrowtail M_\alpha \twoheadrightarrow H_\alpha$ as usual denote the representing sequence for the functor p^α. Form the exact sequence $\sum Z \rightarrowtail \sum M_\alpha \twoheadrightarrow \sum H_\alpha$ by taking an appropriate direct sum of copies of e_α so that $F \simeq \sum Z$, say ψ: $F \rightarrowtail\!\!\!\!\twoheadrightarrow \sum Z$. Since $p^\alpha M_\alpha = Z$, we have that $p^\alpha(\sum M_\alpha) = Z$. Hence, there is a diagram

$$\begin{array}{ccccc} F & \rightarrowtail & G & \twoheadrightarrow & G/F \\ \psi \downarrow & & & & \\ p^\alpha(\sum M_\alpha) & \rightarrowtail & \sum M_\alpha & & \end{array}$$

Quoting Theorem 159 yields a homomorphism ϕ: $G \to \sum M_\alpha$ which makes the above diagram commutative. Since $F \cap \operatorname{Ker}\phi \subseteq \operatorname{Ker}\psi = 0$, it follows that $\operatorname{Ker}\phi = 0$ since F is necessarily an essential subgroup of G. Thus, ϕ is an embedding of G into $\sum M_\alpha$. By Lemma 171, G is free.

Corollary 173. Each torsion free weak p^α-projective is free.

Proof. This statement is an immediate consequence of Theorem 165 and Theorem 172.

Corollary 174. If a torsion free group G contains a free subgroup F such that G/F is contained in a sum of countable reduced p-groups, then G is free.

Our insistence that only one prime be allowed rests upon the fact that there is an immediate counterexample when infinitely many primes are used. For example, for N an infinite subset of the primes, there is a noncyclic group B such that $Z \subseteq B \subseteq Q$ and $B/Z \simeq \sum_{p \in N} Z(p)$. However, it is clear the above results remain true when G/F has a finite number of primary components and each of these components is an appropriate weak projective. Although it remains open (at the time of this writing) as to whether G is necessarily free when G/F is only assumed to be a reduced p-group, we can make some comment on the general situation.

<u>Theorem 175</u>. If a torsion free group G is an extension of free group F by a reduced p-group, then G is $\aleph_1$-free and slender.

<u>Proof</u>. If C is a countable subgroup of G, then applying Corollary 174 to the exact sequence $C \cap F \rightarrowtail C \twoheadrightarrow \{C,F\}/F \subseteq G/F$ shows that C is free. Hence, G is $\aleph_1$-free. To show that an $\aleph_1$-free group is slender, it suffices by Corollary 156 to show that the Specker group P cannot be isomorphic to a subgroup of the group. So suppose that P is a subgroup of G. From the exact sequence $F \cap P \rightarrowtail P \twoheadrightarrow \{P,F\}/F \subseteq G/F$ and from Theorem 153, we obtain that {P,F}/F is bounded which further implies the contradictory fact that P is free.

Our next considerations in this chapter concern the splitting of mixed groups, that is, groups G such that $G \simeq tG + (G/tG)$. Indeed, we determine the structure of those torsion groups T such that G is a splitting mixed group whenever $T \simeq tG$ and, likewise, we find the strucutre of those torsion free groups A such that G is a splitting mixed group whenever $A \simeq G/tG$. Baer was the chief pioneer in this part of abelian group theory and, in fact, the solution of the first of the above problems is contained in [9]. Also, Fomin [31] solved this problem about the same time. Finally, the author [43] solved the second of the above problems.

Theorem 176. (Baer-Fomin) A torsion group T is a direct summand of every group G for which $T = tG$ if and only if $G = B + D$ where B is bounded and D is torsion divisible.

Proof. From Theorem 12 and Theorem 19, we deduce the sufficiency of the above result. To prove the necessity, it, of course, suffices to assume that T is reduced and show that T is necessarily bounded. To this end we observe that the exact sequence $T \rightarrowtail E(T) \twoheadrightarrow D$ splits, where $T \rightarrowtail E(T)$ is the cannonical embedding of T into its cotorsion completion $E(T) \simeq Ext(Q/Z,T)$ (recall that D is torsion free and divisible). By Lemma 45, it follows that both T and T/T' are cotorsion ($T' = \bigcap nT$). Let $H = T/T'$ and consider the pure exact sequence $H \rightarrowtail \hat{H} \twoheadrightarrow M$ where $\hat{H}$ is the n-adic completion of H and M is divisible. If $K \supseteq H$ such that $K/H = M/tM$, then $K \simeq H + (K/H)$ since $Ext(K/H,H) = 0$. Since $\hat{H}$ is reduced, it follows that $M = tM$ and that $\hat{H}$ is a torsion group. By Exercise 20 (III), $\hat{H}$ and, hence, H is bounded, that is, T/T' is a bounded group. We finally call upon Exercise 5 to show that T itself is bounded.

Corollary 177. A reduced torsion group T is cotorsion if and only if it is bounded.

Before giving the author's result concerning the aforementioned problems, we need a lemma which is of some independent interest itself.

Lemma 178. Given any torsion free group G, there is a torsion T and an extension M of T by G such that each torsion free subgroup of M is free.

Proof. We first prove the above lemma when $G = Q$.

For each prime p, let $A^p = \prod_{n<\omega}\{b_{pn}\}$ where $\{b_{pn}\} \simeq Z(p^n)$ and set $A = \prod_p A^p$. Let $u = \langle u_p \rangle \in A$ where $u_p \in A^p$ is defined as follows: $u_p = \langle p^{e_n} b_{pn} \rangle$ where $e_n = [n/2]$ ([ ] denotes the greatest integer function). Note that $h_p^{A_p}(u_p) = 0$ since $e_1 = 0$. It is elementary to see that u_p is an element of infinite order in A^p and, hence, that u is an element of infinite order in A. It is also easily seen, for each $y = \langle y_p \rangle \in tA$ and each prime p, that $h_p^A(nu + y) = h_p^{A_p}(nu_p + y_p) < \infty$ where n is a nonzero integer.

To show that $h_p^{(A/tA)}(u + tA) = \infty$ for each p, it suffices to show that $u_p - x_p^{(m)} \in p^m A^p$ for some $x_p^{(m)} \in A^p$ and any $m > 0$. Clearly, there is a positive integer N such that if $n \geq N$, then $e_n \geq m$. Define $x_p^{(m)} = \langle \delta_n b_{pn} \rangle$ where $\delta_n = p^{e_n}$ for $n = 1,\ldots,N$ and $\delta_n = 0$ for $n > N$. Then $u_p - x_p^{(m)} \in p^m A^p$. Since A/tA is torsion free, there is a pure subgroup H of A containing tA and u such that H/tH = H/tA is a pure rank one subgroup of A/tA. Since rank(H/tH) = 1 and $h_p^{(H/tH)}(u + tH) = \infty$ for each p, it follows that $H/tH \simeq Q$.

Suppose that C is a subgroup of H such that $C \cap tH = 0$. We may suppose that $C \neq 0$. Since $C \simeq \{C,tH\}/tH \subseteq H/tH \simeq Q$, it follows that C has rank one. Let $c \neq 0 \in C$. Since H/tH = Q, there are nonzero integers r and s such that rc = su + y, where $y = \langle y_p \rangle \in tH$. We have already observed that $h_p^H(rc) = h_p^H(su + y) = h_p^A(su + y) < \infty$ for each p. Hence, $h_p^C(rc) < \infty$ for each prime p. There is a positive integer N such that $y_p = 0$ for all $p > N$. Let p be any prime larger than $N + |s|$. Then $(s,p) = 1$ and $h_p^C(rc) \leq h_p^H(su + y) = h_p^{A^p}(su_p) = h_p^{A^p}(u_p) = 0$. Hence, the height sequence for rc in C is zero except for a finite number of primes and contains no "infinities". By Theorem 116, $C \simeq Z$.

Now let G be an arbitrary torsion free group and let $D(G) = \sum_m Q$, where as always D(G) is the injective envelope of G. By taking a direct sum of m copies of the exact sequence $tH \rightarrowtail H \twoheadrightarrow Q$, we obtain an exact sequence $T \rightarrowtail N \twoheadrightarrow D(G)$, where $T = \sum_m tH$ and $N = \sum_m H$. Let M be the subgroup of N containing T such that $M/T = G \subseteq D(G)$. If F is a torsion free subgroup of M (hence, F is also a torsion free subgroup of $N = \sum_m H$), it follows from the above paragraph and Theorem 170 that F is free.

A rather immediate consequence of the above lemma is the following...

Theorem 179. A torsion free group G has the property that every mixed group M splits whenever $M/tM \simeq G$ if and only if G is free.

Proof. Since the sufficiency is clear, we proceed straight to the case of the necessity. So suppose that Ext(G,T) = 0 for every torsion group T. We apply Lemma 178 and obtain an extension $T \rightarrowtail M \twoheadrightarrow G$ where each torsion free subgroup of M is free. Since $T \rightarrowtail M \twoheadrightarrow G$ splits by hypothesis, G is necessarily free.

At this point one should observe that the above result holds for modules over a principal ideal domain. In particular, the above theorem holds for modules over the ring I_N where N is a subset of the primes and $I_N = \{n/m \in Q: (m,p) = 1 \text{ for } p \in N\}$. Thus, I_N is just the localization of the integers to the set of primes N. We also use the symbol I_N to denote the additive group of I_N; however, no confusion should arise. As was the case for I_p $(N = [p])$ in Chapter III, a torsion group T with $T_p = 0$ for $p \in N$ can be considered a module in a natural way over I_N. In particular, $I_N \otimes G$ is naturally a module over I_N for any group G. Moreover, one should note that $\mathrm{Hom}(I_N \otimes G, T) = \mathrm{Hom}_{I_N}(I_N \otimes G, T)$ when $T_p = 0$ for $p \in N$.[3] We now prove a technical lemma and then proceed to push our above results slightly further.

Lemma 180. Let N be a nonempty subset of the primes and let T be a torsion group such that $T_p = 0$ for $p \notin N$. If G is a group such that $\mathrm{Tor}(G, I_N/Z) = 0$, then $\mathrm{Ext}(G,T)$, $\mathrm{Ext}(I_N \otimes G, T)$, and $\mathrm{Ext}_{I_N}(I_N \otimes G, T)$ are isomorphic as abelian groups.

Proof. We may assume that N is a proper subset of the primes; since otherwise $I_N = Z$. Let $\tilde{N}$ be the set of primes not in N. From the definition of I_N, we obtain the exact sequence $Z \rightarrowtail I_N \twoheadrightarrow \sum_{p \in \tilde{N}} Z(p)$ which yields the exact sequence $Z \otimes G \rightarrowtail I_N \otimes G \twoheadrightarrow \sum_{p \in \tilde{N}} Z(p^\infty) \otimes G$, since $\mathrm{Tor}(I_N/Z, G) = 0$. Hence, we obtain a second exact sequence $\mathrm{Ext}(\sum_{p \in \tilde{N}} Z(p^\infty) \otimes G, T) \rightarrowtail \mathrm{Ext}(I_N \otimes G, T) \twoheadrightarrow \mathrm{Ext}(Z \otimes G, T)$. Since $\sum_{p \in \tilde{N}} Z(p^\infty) \otimes G$ and T are torsion groups with no nonzero primary components in common, we have that $\mathrm{Ext}(\sum_{p \in \tilde{N}} Z(p^\infty) \otimes G, T) = 0$ and thus $\mathrm{Ext}(I_N \otimes G, T) \simeq \mathrm{Ext}(Z \otimes G, T) \simeq \mathrm{Ext}(G,T)$.

Let E be the I_N-injective envelope (= the Z-injective envelope) of T. The exactness of $T \rightarrowtail E \twoheadrightarrow E/T$ induces exactness of the rows of the

[3] We drop the subscript "R" on the functors $\mathrm{Hom}_R(A,B)$ and $\mathrm{Ext}_R(A,B)$ only when R = Z.

commutative diagram

$$\begin{array}{ccccccc}
\mathrm{Hom}(I_N \otimes G,T) & \longrightarrow & \mathrm{Hom}(I_N \otimes G,E) & \longrightarrow & \mathrm{Hom}(I_N \otimes G,E/T) & \twoheadrightarrow & \mathrm{Ext}(I_N \otimes G,T) \\
\| & & \| & & \| & & \\
\mathrm{Hom}_{I_N}(I_N \otimes G,T) & \longrightarrow & \mathrm{Hom}_{I_N}(I_N \otimes G,E) & \longrightarrow & \mathrm{Hom}_{I_N}(I_N \otimes G,E/T) & \twoheadrightarrow & \mathrm{Ext}_{I_N}(I_N \otimes G,T)
\end{array}$$

Thus, the cokernels are isomorphic.

Definition. For a nonempty subset N of the primes, we call G a B^N-group provided Ext(G,T) = 0 for each torsion group T with $T_p = 0$ for $p \in N$. Since subgroups of B^N-groups are again B^N-groups (Exercise 7) and since Z(p) is clearly not a B^N-group for $p \in N$, it follows from Theorem 179 that, for N the set of all primes, G is a B^N-group if and only if G is free.

Theorem 181. A group G is a B^N-group if and only if $I_N \otimes G$ is free as an I_N-module.

Proof. Suppose first that G is a B^N-group. Then from our above discussion, $tG_p = 0$ for $p \in N$. Therefore, $I_N \otimes tG = 0$ and, hence, $I_N \otimes G \simeq I_N \otimes (G/tG)$. Let T be a torsion group such that $T_p = 0$ for $p \in N$. We have an exact sequence $0 = \mathrm{Hom}(tG,T) \to \mathrm{Ext}(G/tG,T) \to \mathrm{Ext}(G,T) = 0$ which implies that G/tG is a B^N-group. An application of Lemma 180 shows that $\mathrm{Ext}_{I_N}(I_N \otimes G,T) = 0$ for each torsion I_N-module. The remarks following Theorem 179 show that $I_N \otimes G$ is free as an I_N-module, that is, $I_N \otimes G = \sum_m I_N$. Finally, an application of Lemma 180 also easily settles the sufficiency of the above result.

The structure of B^N-groups is contained in...

Theorem 182. G is a B^N-group if and only if $tG_p = 0$ for each $p \in N$ and G/tG is isomorphic to a subgroup of $\sum_m I_N$ where m = rank(G/tG).

Proof. Suppose first that G is a B^N-group. By Theorem 181, $I_N \otimes G$ is a free I_N-module. Hence, $I_N \otimes tG = 0$ which implies that tG = 0 for each $p \in N$. Hence, $I_N \otimes G \simeq I_N \otimes (G/tG)$. The exact sequence $Z \rightarrowtail I_N \twoheadrightarrow I_N/Z$ induces the exact sequence

$G/tG \simeq Z \otimes (G/tG) \rightarrowtail I_N \otimes (G/tG) \twoheadrightarrow (I_N/Z) \otimes (G/tG)$. Since $(I_N/Z) \otimes (G/tG)$ is torsion, it follows that $\text{rank}(G/tG) = \text{rank}(I_N \otimes G)$. This completes the proof of the necessity.

Suppose that $tG_p = 0$ for each $p \in N$ and that $G/tG \subseteq \sum_m I_N$. Then $I_N \otimes tG = 0$ and $I_N \otimes G \simeq I_N \otimes (G/tG) \subseteq \sum_m I_N \otimes I_N \subseteq \sum_m I_N$. Hence, $I_N \otimes G$ is an I_N-submodule of a free I_N-module and thus $I_N \otimes G$ is a free I_N-module. We finish the proof by quoting Theorem 181.

Let N be a nonempty proper subset of the primes. Although Theorem 182 implies there are nonfree B^N-groups (for example, the group I_N), one might suspect that separable B^N-groups are free. However, we shall presently show, for each nonempty proper subset N of the primes, that there is a pure subgroup of the Specker group P and, hence, a separable group which is a nonfree B^N-group.

Proposition 183. Let N be a nonempty proper subset of the primes. Then there is a separable B^N-group that is not free.

Proof. We invoke Theorem 146 to obtain a pure subgroup A of the Specker group such that $A = \bigcup_{\alpha<\Omega} A_\alpha$ with

(1) $A_\alpha \simeq \sum_{\aleph_0} Z$ for each α.

(2) $A_\alpha = \bigcup_{\beta<\alpha} A_\beta$ if β is a limit ordinal.

(3) $A_{\alpha+1}/A_\alpha \simeq I_N$ for each $\alpha < \Omega$.

Theorem 146 also guarantees that A is not free since I_N is not free. Now observe that $I_N \otimes A = \bigcup_{\alpha<\Omega}(I_N \otimes A_\alpha)$ and that

(1') $I_N \otimes A_\alpha \simeq \sum_{\aleph_0} I_N$.

(2') $I_N \otimes A_\alpha = \bigcup_{\beta<\alpha}(I_N \otimes A_\beta)$ if β is a limit ordinal.

(3') $I_N \otimes A_{\alpha+1}/I_N \otimes A_\alpha \simeq I_N \otimes I_N \simeq I_N$.

Since I_N is projective as an I_N-module, it follows that $I_N \otimes A_{\alpha+1} = I_N \otimes A_\alpha + C_\alpha$ where $C_\alpha \simeq I_N$ for each α. Thus, $I_N \otimes A$ is free as an I_N-module and so, by Theorem 181, A is a B^N-group.

Another curious example is found in...

Proposition 184. (Chase) There is a nonfree, separable group G such that $\mathrm{Ext}(G,T_p) = 0$ for each p-group T_p and for each prime p.

Proof. Let I be a rank one torsion free group of type $[(1,1,1,\ldots,1,\ldots)]$. We again apply Theorem 146 to obtain a nonfree, pure subgroup $A = \bigcup_{\alpha<\Omega} A_\alpha$ of the Specker group such that

(1) $A_\alpha \simeq \sum_{\aleph_0} Z$ for each α.

(2) $A_\alpha = \bigcup_{\beta<\alpha} A_\beta$ if β is a limit ordinal.

(3) $A_{\alpha+1}/A_\alpha \simeq I$ for each α.

The remainder of the proof goes like that of Proposition 183 after one observes that $I \otimes I_p \simeq I_p$ for each prime p.

One may also show that the above group A has the property that $\mathrm{Ext}(A,T)$ is torsion free for all torsion groups T.

Our concluding considerations in this chapter have to do with a problem attributed to J. H. C. Whitehead. Whitehead's Problem asks for a characterization of those groups G for which $\mathrm{Ext}(G,Z) = 0$. A solution to this problem is not known (at this time). However, we shall present some of the known structure of such groups.

Definition. A group G is called a W-group if it satisfies $\mathrm{Ext}(G,Z) = 0$.

Proposition 185. The class of W-groups is closed under the operations of taking arbitrary direct sums and of taking subgroups.

Proof. If $[G_i]_{i\in I}$ is a family of W-groups and if $G = \sum_{i\in I} G_i$, then $\mathrm{Ext}(G,Z) = \mathrm{Ext}(\sum_{i\in I} G_i, Z) = \prod_i \mathrm{Ext}(G_i,Z) = 0$ which implies that G is also a W-group.

Let H be a subgroup of the W-group G. The exactness of $0 = \mathrm{Ext}(G,Z) \twoheadrightarrow \mathrm{Ext}(H,Z)$ shows that H is also a W-group.

Theorem 186. (Stein) W-groups are $\aleph_1$-free.

Proof. Let G be a W-group, $G \neq 0$. The exact sequence $Z \rightarrowtail \hat{Z} \twoheadrightarrow D$ induces the exact sequence $\mathrm{Hom}(G,Z) \rightarrowtail \mathrm{Hom}(G,\hat{Z}) \rightarrow \mathrm{Hom}(G,D) \rightarrow \mathrm{Ext}(G,Z) = 0$. If $\mathrm{Hom}(G,Z) = 0$, then $\mathrm{Hom}(G,\hat{Z}) \simeq \mathrm{Hom}(G,D)$ which is impossible since $\mathrm{Hom}(G,D)$ is nonzero and divisible and $\mathrm{Hom}(G,\hat{Z})$ is reduced (because $\hat{Z}$ is reduced). Hence, Hom(G,Z) ≠ 0 if G is a nonzero W-group. This shows that G ≃ H + Z for some H ⊆ G. Thus, if G has finite rank, we may apply Proposition 185 and induction to show that G is free. It follows that the subgroups of finite rank in W-groups are necessarily free. By Pontryagin's Theorem (137), a W-group is $\aleph_1$-free.

Theorem 187. (Rotman) W-groups are separable.

Proof. Let G be a W-group and let A be a pure finitely generated subgroup of G. By Theorem 186, A is free of finite rank and so A ≃ Hom(A,Z). Therefore, we obtain an exact sequence A ≃ Hom(A,Z) → Ext(G/A,Z) → Ext(G,Z) = 0, that is, Ext(G/A,Z) is an epimorphic image of A and so is finitely generated. However, since G/A is torsion free, Ext(G/A,Z) is divisible (Ext(G/A,Z) is an epimorphic image of Ext(D(G/A),Z)). Hence, Ext(G/A,Z) = 0 which implies that $\mathrm{Ext}(G/A,\sum_m Z) = \prod_m \mathrm{Ext}(G/A,Z) = 0$, where $A \simeq \sum_m Z$. (Recall that m is finite!) Therefore, Ext(G/A,A) = 0 and so $A \rightarrowtail G \twoheadrightarrow G/A$ splits.

Proposition 188. (Rotman-Nunke) The Specker group is not a W-group.

Proof. We first show that $t\mathrm{Ext}(\hat{Z},Z) \neq 0$. Now $Z \rightarrowtail \hat{Z} \rightarrowtail D \simeq \sum_{2^{\aleph_0}} Q$ induces the exact sequence $Z \simeq \mathrm{Hom}(Z,Z) \rightarrowtail \mathrm{Ext}(D,Z) \twoheadrightarrow \mathrm{Ext}(\hat{Z},Z)$. Since $Z \not\simeq \mathrm{Ext}(D,Z)$, it follows that $t\mathrm{Ext}(\hat{Z},Z) \simeq Q/Z$. Let P denote the Specker group and let S be the subgroup of finitely nonzero sequences in P. We then obtain an exact sequence

$\mathrm{Hom}(P,Z) \rightarrowtail \mathrm{Hom}(S,Z) \xrightarrow{\delta} \mathrm{Ext}(P/S,Z) \twoheadrightarrow \mathrm{Ext}(P,Z)$. From Exercise 20 (Chapter IV), $\hat{Z} \simeq \prod_p I_p^*$ is isomorphic to a direct summand of P/S. Therefore, $t\mathrm{Ext}(P/S,Z) \neq 0$. By Proposition 148, the above embedding $\mathrm{Hom}(P,Z) \rightarrowtail \mathrm{Hom}(S,Z)$ is a pure embedding. Hence, Imageδ is a torsion free subgroup of Ext(P/S,Z) and so $t\mathrm{Ext}(P,Z) \neq 0$. Thus, Ext(P,Z) ≠ 0.

Corollary 189. (Rotman-Nunke) W-groups are slender.

The following theorem shows that a significant increase in the amount of information can be gained if we assume that Ext(G,F) = 0 for some free group of infinite rank. However, we still have no conclusive results.

Theorem 190. Let F be a free group of infinite rank α. Then Ext(G,F) = 0 if and only if the following conditions are satisfied:

(i) Every subgroup of G of cardinality less than or equal to α is free.

(ii) Every subgroup of G of index α contains a direct summand of G of index α.

Proof. Suppose that G satisfies (i) and (ii) above and that $F \overset{i}{\rightarrowtail} M \overset{j}{\twoheadrightarrow} G$ represents an element of Ext(G,F) where i is assumed to be the inclusion homomorphism. (Clearly, we also may assume that $|G| \geq \alpha$.) Let H be a subgroup of M maximal with respect to the property that $F \cap H = 0$. Hence, $H \simeq j(H)$ and $|G/j(H)| \leq \alpha$ because $\alpha \geq \aleph_0$ and because M/H is an essential extension of $F \simeq (F + H)/H$, and so both have the same injective envelope. Therefore, $G = K + B$ where $K \subseteq j(H)$ and $B \simeq F$ by property (ii). Let $C = j^{-1}(B)$ and let $E = j^{-1}(K) \cap H$. If $x \in M$, then $j(x) = k + b$ where $k \in K$ and $b \in B$. Since $K \subseteq j(H)$, we have that $j^{-1}(K) \subseteq H + F$. Therefore, there are elements $e \in E$ and $c \in C$ such that $j(e) = k$ and $j(c) = b$. Since $j(x - (e + c)) = 0$, then $x - (e + c) \in F \subseteq C$, which implies that $x \in \{E,C\}$. Thus, $M = \{E,C\}$. If $x \in E \cap C$, then $j(x) \in B \cap K = 0$, which implies $x \in F \cap E \subseteq F \cap H = 0$. Hence, $M = E + C$ where $F \subseteq C$ and $|C| \leq \alpha$. But C/F is isomorphic to a subgroup of G of cardinality not exceeding α; hence, C/F is free. It follows that F is a direct summand of M, that is, $F \overset{i}{\rightarrowtail} M \overset{j}{\twoheadrightarrow} G$ splits. Thus, Ext(G,F) = 0.

Now suppose that Ext(G,F) = 0 and suppose that H is a subgroup of G such that $|H| \leq \alpha$. The induced epimorphism $0 = \text{Ext}(G,F) \twoheadrightarrow \text{Ext}(H,F)$ shows that Ext(H,F) = 0. Moreover, $|H| \leq \alpha$ implies that H has a free resolution $F_0 \rightarrowtail F \twoheadrightarrow H$ where $|F_0| \leq \alpha$. Hence, $\text{Ext}(H,F_0) = 0$ and so $F_0 \rightarrowtail F \twoheadrightarrow H$ splits which yields that H is free. Thus, (i) is satisfied.

Assume that H is a subgroup of G of index α. Then there is a subgroup A of G of cardinality less than or equal to α such that $G = \{A,H\}$. The preceding paragraph gives that A is free. We form the short exact sequence

$C \overset{i}{\rightarrowtail} A + H \overset{\sigma}{\twoheadrightarrow} G$ where A + H is the outer direct sum of A and H and $\sigma: (a,h) \to a + h$. Clearly, $C = \text{Ker}\sigma = \{(x,-x): x \in A \cap H\}$ is isomorphic to a free subgroup of A of cardinality less than or equal to α. Since necessarily Ext(G,C), it follows that there is a homomorphism $\theta: G \to A + H$ such that $\sigma\theta = 1_G$. Let π be the natural projection of A + H onto A and define $\psi: G \to A$ by $\psi = \pi\theta$. Since A is a free group with $|A| \leq \alpha$, $G = \text{Ker}\psi + B$ where B is a free group such that $|B| \leq \alpha$. It remains only to show that $\text{Ker}\psi \subseteq H$. Now $g \in \text{Ker}\psi$ if and only if $\theta(g) = (0,h) \in A + H$. But this implies that $g = \sigma\theta(g) = \sigma(0,h) = h \in H$.

The above result leads us to the following definition.

Definition. A group G is called coseparable ($\aleph_1$-coseparable) if G is $\aleph_1$-free and if every subgroup H of G with the property that G/H is finitely (countably) generated contains a direct summand K of G such that G/K is finitely (countably) generated. Compare these definitions with those of separability and $\aleph_1$-separability in Chapter VII. We further refer to a group G as totally separable (totally $\aleph_1$-separable) provided every subgroup of G is separable ($\aleph_1$-separable). The discussion following Theorem 140 shows that the Specker group is not totally separable (and obviously not totally $\aleph_1$-separable).

An immediate consequence of Theorem 190 and the above definitions is our next result.

Theorem 191. A group G is $\aleph_1$-coseparable if and only if $\text{Ext}(G,\sum_{\aleph_0} Z) = 0$.

Although the property of being $\aleph_1$-coseparable is on the surface a weaker property than that of being free, the problem remains open as to whether the two properties are or are not equivalent. Along this same direction, Chase [18] has shown that a group G is coseparable if and only if Ext(G,Z) is torsion free. With the assumption of the Continuum Hypothesis, Chase [17] provides an example of a nonfree coseparable group.

The reason for considering totally separable groups and totally

$\aleph_1$-separable groups is contained in our next two results.

Theorem 192. A totally separable group is coseparable.

Proof. Let G be a totally separable group and suppose that H is a subgroup of G such that G/H is finitely generated. Then G = {C,H} where C is finitely generated; hence, C is free of finite rank. Since H is also separable, H = K + B where B has finite rank and where $C \cap H \subseteq B$. Let A = {B,C}. If $k \in K \cap A$, then k = b + c where $b \in B$ and $c \in C$. Hence, $c = k - b \in H \cap C \subseteq B$. But this implies that $k \in B \cap K = 0$. Observing that G = {K,A}, we have that G = K + A, $K \subseteq H$ and that G/K is finitely generated. Thus, G is coseparable.

Examination of this proof shows that, with obvious changes, one obtains...

Theorem 193. A totally $\aleph_1$-separable group is $\aleph_1$-coseparable.

This result adds possibly more significance to Theorem 147 which yields the existence of a group G that is nonfree and $\aleph_1$-separable. Is G totaly $\aleph_1$-separable? We next consider a partial converse of Theorem 193.

Theorem 194. If $\mathrm{Ext}(G,\sum_{\aleph_0} Z) = 0$ and if G is $\aleph_1$-separable, then G is totally $\aleph_1$-separable.

Proof. Let H be a subgroup of G and let A be a countable subgroup of H. By hypothesis, G = B + K where B is a countable free subgroup of G containing A. Let π be the natural projection of G onto K restricted to H. Therefore, $B \cap H \xrightarrow{i} H \xrightarrow{\pi} \pi(H)$ is exact and $B \cap H$ is isomorphic to a direct summand of $S = \sum_{\aleph_0} Z$. For any subgroup C of G, $\mathrm{Ext}(C,\sum_{\aleph_0} Z) = 0$ since there is an epimorphism $0 = \mathrm{Ext}(G,\sum_{\aleph_0} Z) \longrightarrow \mathrm{Ext}(C,\sum_{\aleph_0} Z)$. Hence, $\mathrm{Ext}(\pi(H),\sum_{\aleph_0} Z) = 0$ and, therefore, $\mathrm{Ext}(\pi(H),B \cap H) = 0$. Hence, $B \cap H \xrightarrow{i} H \xrightarrow{\pi} \pi(H)$ splits and thus $B \cap H$ is a countable direct summand of H containing A.

One may use the above theorem to reduce the amount of work in showing that a group G is totally $\aleph_1$-separable.

Theorem 195. If a subgroup H of a group G is $\aleph_1$-separable whenever

G/H is countable, then G is totally $\aleph_1$-separable.

Proof. The proofs of Theorem 190 and Theorem 193 show that our present hypothesis is all that is demanded in establishing that $\text{Ext}(G,\sum_{\aleph_0} Z) = 0$. Hence, by Theorem 194, G is totally $\aleph_1$-separable.

It is elementary to show that an arbitrary direct sum of totally $\aleph_1$-separable groups is totally $\aleph_1$-separable. We prove a stronger result.

Theorem 196. If G and H are totally $\aleph_1$-separable, then for any extension $H \xrightarrow{i} M \xrightarrow{j} G$ of H by G, M is also totally $\aleph_1$-separable.

Proof. We may assume that i is the inclusion homomorphism. Since there is an exact sequence $0 = \text{Ext}(G,\sum_{\aleph_0} Z) \rightarrowtail \text{Ext}(M,\sum_{\aleph_0} Z) \twoheadrightarrow \text{Ext}(H,\sum_{\aleph_0} Z) = 0$, it follows that $\text{Ext}(M,\sum_{\aleph_0} Z) = 0$. By Theorem 195, it is enough to show that M is $\aleph_1$-separable. Therefore, let A be a countable subgroup of M. Since $A/A \cap H$ is isomorphic to a subgroup of G and since $|A/A \cap H| \leq \aleph_0$ we have that $A = A \cap H + A_0$. Since H is totally $\aleph_1$-separable, $H = H_0 + H_1$ where H_0 is countable and contains $A \cap H$. Let K be a subgroup of M which is maximal with respect to the property that $K \cap \{H,A_0\} = 0$. It follows that $|M/(H_1 + K)| \leq \aleph_0$. Hence, there is a countable free subgroup C of M containing $\{H_0,A_0\}$ such that $M = \{C,H_1 + K\}$. Since both H_1 and K are $\aleph_1$-separable, it is easily seen that $H_1 + K$ is $\aleph_1$-separable. Hence, $H_1 + K = Y + X$ where Y is countable and contains $C \cap (H_1 + K)$. Set $F = \{C,Y\}$. It is straightforward to check that $M = F + X$ and that F is countable. Moreover, $F = \{C,Y\} \supseteq \{H_0,A_0\} \supseteq A$.

Although we cannot show that $\text{Ext}(G,\sum_{\aleph_0} Z) = 0$ implies that G is $\aleph_1$-separable, we can show that $\text{Ext}(G,\sum_{\aleph_0} Z) = 0$ implies a slightly weaker statement.

Theorem 197. If $\text{Ext}(G,\sum_{\aleph_0} Z) = 0$ and if A is a countable subgroup of G, then there is a decomposition $G = B + H$ where B is countable and $A \cap H = 0$.

Proof. Let K be a subgroup of G maximal with respect to the property that $A \cap K = 0$. Hence, $|G/K| \leq \aleph_0$ which implies by Theorem 191 that $G = H + B$ where $H \subseteq K$ and $|B| \leq \aleph_0$. Since $A \cap H \subseteq A \cap K = 0$, the proof is complete.

Corollary 198. If $\operatorname{Ext}(G,\sum_{\aleph_0} Z) = 0$ and if rank $G \geq \aleph_0$, then G contains a free direct summand of rank $\aleph_0$.

Proof. Using the notation of Theorem 197, choose A such that rank $A = \aleph_0$. Since $A \cap H = 0$, we have that $\aleph_0 =$ rank $A \leq$ rank B. However, $|B| \leq \aleph_0$ implies that rank $B = \aleph_0$.

We end this chapter with several structural results on $\aleph_1$-coseparable groups. These results exhibit rather strong properties for such groups. We use the symbol S to denote the group $\sum_{\aleph_0} Z$.

Theorem 199. If G is an $\aleph_1$-coseparable group and if H is a subgroup of G such that G/H is countable, then $S + H \simeq S + G$.

Proof. If rank $G \leq \aleph_0$, then G is free and the conclusion follows. Hence, suppose that rank $G > \aleph_0$. Since $|G/H| \leq \aleph_0$, there is a countable free subgroup A of G such that rank $A = \operatorname{rank}(A \cap H) = \aleph_0$ and such that $G = \{A,H\}$. We now use the fact that the short exact sequence $\operatorname{Ker}\sigma \rightarrowtail A + H \xrightarrow{\sigma} G$ (defined in the proof of Theorem 190) splits. The proof is complete when one notes that $S \simeq A \simeq \operatorname{Ker}\sigma$.

Corollary 200. If the $\aleph_1$-coseparable group G has infinite rank, then $G \simeq G + S$.

Proof. By Corollary 198, G contains a direct summand H such that $G \simeq H + S$ and, by Theorem 199, $H + S \simeq G + S$.

Theorem 201. If G is an uncountable $\aleph_1$-coseparable group, then $G \simeq H$ for each subgroup H satisfying G/H countable.

Proof. Observing that H is also an uncountable group satisfying $\operatorname{Ext}(H,S) = 0$, we apply Theorem 199 and Corollary 200 to obtain $H \simeq S + H \simeq S + G \simeq G$.

Exercises

1. Prove Lemma 158.

2. A p^α-projective is a weak p^α-projective.

3. Let $u_1 = \bigcap_p p^\omega$. Show that u_1 has $e = \langle e_{\omega,p}\rangle \in \prod_p \mathrm{Ext}(H_{\omega,p}, Z)$ as its representing sequence, where $e_{\omega,p}: Z \rightarrowtail M_{\omega,p} \twoheadrightarrow H_{\omega,p}$ represents p^ω. (Note e has the form $e: Z \rightarrowtail M \twoheadrightarrow \sum_p H_{\omega,p}$.) Prove that the u_1-projectives are just the pure projectives.

4. Check theorems 158, 159, 161, 165 for the functor u_1 (defined in Exercise 3).

5. If $G' = \bigcap nG$ and if G/G' is a bounded group, then G itself is bounded.

6. Show that Theorem 170 remains true if the groups M_i, $i \in I$, are allowed to have countable torsion free rank.

7. Prove that the class of B^N-groups, for fixed N, is closed under the operations of taking arbitrary direct sums, extensions and of taking subgroups.

8. Prove that if G is a B^{N_i}-group for $i = 1,\ldots,n$, then G is a B^N-group where $N = \bigcup_{i=1}^{n} N_i$ (n is finite). Show that this statement is, in general, false if we allow infinitely many N_i.

9. Prove that $I_N \otimes I_N \simeq I_N$ and that $I \otimes I_p \simeq I_p$ where I is a rank one torsion free group of type $[(1,1,1,1,\ldots,1,\ldots)]$.

10. Let N be a novoid subset of the primes. If A is a group such that $tA_p = 0$ for $p \in N$ and if every torsion free subgroup of A is free, then A is a B^N-group. (<u>Hint</u>: Choose F maximal with respect to $F \cap tA = 0$ and show that $I_N \otimes A \simeq I_N \otimes F$.)

11. If A is as in Exercise 10, then A/tA is also a B^N-group.

12. Call M a direct sum of p-mixed groups if $M = \sum_p M_p$, where tM_p is p-primary. If N consists of all primes except the prime p, we use the notation B^p instead of B^N. If the torsion free group $G = \sum_p G_p$, where G_p is a

B^p-group for each prime p, then any extension of a torsion group T by G is a direct sum of p-mixed groups.

13. Prove the converse of Exercise 12, that is, if every extension of a torsion group T (T arbitrary) by the torsion free group G is a direct sum of p-mixed groups, then $G = \sum_p G_p$ where G_p is a B^p-group for each prime p.

14. Prove that if Ext(G,F) = 0 for each free group F, then G itself is a free group.

15. Show that if Ext(G,Z) is torsion free and if G is reduced, then G is separable. (<u>Hint</u>: For A a finitely generated pure subgroup of G, consider the exact sequence $\mathrm{Hom}(G,Z) \longrightarrow \mathrm{Hom}(A,A) \longrightarrow \mathrm{Ext}(G/A,A) \twoheadrightarrow \mathrm{Ext}(G,A)$.)

16. If G and K are $\aleph_1$-coseparable groups of infinite rank and if $G/A \simeq K/B$, where A and B are countable subgroups of G and K, respectively, then $G \simeq K$.

17. Let G be a reduced nonzero torsion free group. Prove, for some free group F, that Ext(G,F) is neither torsion nor torsion free. (<u>Hint</u>: Consider both the diagram

$$\begin{array}{ccccc} & & & & G \\ & & & & \big\downarrow \\ pF & \rightarrowtail & F & \twoheadrightarrow & F/pF \end{array}$$

where $G/pG \simeq F/pF$ and F is free, and consider a free resolution of G.)

18. Let p be a fixed prime and let $\mathcal{N}_p$ be the class of all torsion free groups which are extensions of free groups by reduced p-groups.

(a) Show that $\mathcal{N}_p$ is closed under the operations of taking arbitrary direct sums and of taking subgroups.

(b) If F is free and G is in $\mathcal{N}_p$, then any extension of F by G is also in $\mathcal{N}_p$.

(c) If each G in $\mathcal{N}_p$ is separable, use (a) and (b) to show that each G in $\mathcal{N}_p$ is a W-group. (Whether or not all groups in $\mathcal{N}_p$ are separable is unknown.)

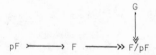

APPENDIX

We wish to present a brief account here of two theorems of Nunke which give rise to the crucial Theorems 76 and 77, and Corollary 78 in Chapter VI, which form the foundation of Nunke's [102], [103], [105] development of p^{α}-purity. We begin with a few preliminary lemmas.

Lemma. For any groups A, B, and C, there is a natural isomorphism ψ: Hom(A $\otimes$ B,C) $\to$ Hom(A,Hom(B,C)) defined, for f ε Hom(A $\otimes$ B,C), by $\psi(f) = g$ where $(g(a))(b) = f(a \otimes b)$.

The proof of this result is well-known and is left to the reader. Next follows a diagram lemma.

Lemma. Given the commutative diagram

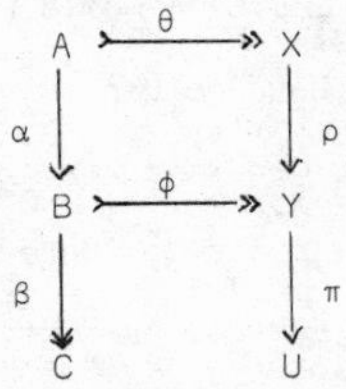

with $\pi\rho = 0$ and the left colume exact, there is a homomorphism $\psi: C \to U$ such that $\psi\beta = \pi\phi$.

Proof. If c ε C, we pick b ε $\beta^{-1}(c)$ and set $\psi(c) = \pi\phi(b)$. To complete the proof, it clearly is enough to show that $\psi(0)$ is necessarily zero in U. So suppose that $\beta(b) = 0$. By exactness of the left column, $\alpha(a) = b$ for some a ε A. Hence, $\phi(b) = \alpha\phi(a) = \rho\theta(a)$ which implies that $\pi\phi(b) = \pi(\rho\theta(a)) = (\pi\rho)\theta(a) = 0$ since $\pi\rho = 0$.

While we do not check commutativity of diagrams in the remainder of our discussion, we remind the reader that all maps in forthcoming statements are induced maps and thus are natural.

Lemma. For groups A, B and C, there is a natural homomorphism

$$\tau\colon \operatorname{Ext}(A \otimes B, C) \to \operatorname{Hom}(A, \operatorname{Ext}(B,C))$$

which is an isomorphism for projective A.

Proof. Let $C \rightarrowtail D \twoheadrightarrow D'$ be an injective resolution of C. From the exact sequence $\operatorname{Hom}(B,D) \rightarrowtail \operatorname{Hom}(B,D') \twoheadrightarrow \operatorname{Ext}(B,C)$ and the natural isomorphism $\operatorname{Hom}(E,\operatorname{Hom}(F,G)) \cong \operatorname{Hom}(E \otimes F, G)$, we obtain the commutative diagram

$$\begin{array}{ccc}
\operatorname{Hom}(A \otimes B, D) & \cong & \operatorname{Hom}(A, \operatorname{Hom}(B,D)) \\
\downarrow & & \downarrow \rho \\
\operatorname{Hom}(A \otimes B, D') & \cong & \operatorname{Hom}(A, \operatorname{Hom}(B,D')) \\
\downarrow & & \downarrow \pi \\
\operatorname{Ext}(A \otimes B, C) & & \operatorname{Hom}(A, \operatorname{Ext}(B,C))
\end{array}$$

with exact left column and $\pi\rho = 0$. The preceding lemma yields a natural homomorphism

$$\psi\colon \operatorname{Ext}(A \otimes B, C) \to \operatorname{Hom}(A, \operatorname{Ext}(B,C)).$$

If A is projective, the right hand column is also exact and, thus, one easily checks that ψ is in fact an isomorphism for projective A.

Lemma. For any groups A, B and C, there is a natural homomorphism

$$\mu'\colon \operatorname{Ext}(A, \operatorname{Ext}(B,C)) \to \operatorname{Ext}(\operatorname{Tor}(A,B), C).$$

Proof. Let $P' \rightarrowtail P \twoheadrightarrow A$ be a projective resolution of A. We then have the induced exact sequence $\operatorname{Tor}(A,B) \rightarrowtail P' \otimes B \longrightarrow P \otimes B$ and the commutative diagram

$$\begin{array}{ccc}
\operatorname{Hom}(P, \operatorname{Ext}(B,C)) & \cong & \operatorname{Ext}(P \otimes B, C) \\
\downarrow & & \downarrow \rho' \\
\operatorname{Hom}(P', \operatorname{Ext}(B,C)) & \cong & \operatorname{Ext}(P' \otimes B, C) \\
\downarrow & & \downarrow \pi' \\
\operatorname{Ext}(A, \operatorname{Ext}(B,C)) & & \operatorname{Ext}(\operatorname{Tor}(A,B), C)
\end{array}$$

with $\pi'\rho' = 0$, exact left column and row isomorphisms obtained from the preceding lemma. Thus, our first lemma provides a (natural) homomorphism

$$\mu': \mathrm{Ext}(A,\mathrm{Ext}(B,C)) \to \mathrm{Ext}(\mathrm{Tor}(A,B),C).$$

If $A' \rightarrowtail A \twoheadrightarrow A''$ is exact and B is any group, there is a natural homomorphism ∂: $\mathrm{Tor}(A'',B) \to A' \otimes B$ which further induces, for any group C, a natural map

$$\mathrm{Ext}(\partial,C): \mathrm{Ext}(A' \otimes B,C) \to \mathrm{Ext}(\mathrm{Tor}(A'',B),C)$$

(explicitly defined in Theorem 76). We now have...

<u>Theorem</u>. (Nunke) The homomorphism μ' is an isomorphism. If $A' \rightarrowtail A \twoheadrightarrow A''$ is an exact sequence of groups inducing a connecting homomorphism ∂: $\mathrm{Tor}(A'',B) \to A' \otimes B$, then the diagram

$$\begin{array}{ccc} \mathrm{Ext}(A' \otimes B,C) & \xrightarrow{\mathrm{Ext}(\partial,C)} & \mathrm{Ext}(\mathrm{Tor}(A'',B),C) \\ \downarrow{\scriptstyle \tau'} & & \downarrow{\scriptstyle \mu'} \\ \mathrm{Hom}(A',\mathrm{Ext}(B,C)) & \xrightarrow{\delta} & \mathrm{Ext}(A'',\mathrm{Ext}(B,C)) \end{array}$$

commutes, where δ is the appropriate connecting homomorphism.

<u>Proof</u>. The commutativity of the diagram follows from the naturality of the maps. To show that μ' is an isomorphism, we let $F_o \rightarrowtail F \twoheadrightarrow A$ be a free resolution of A. We have a commutative diagram

$$\begin{array}{ccccc} \mathrm{Ext}(F \otimes B,C) & \rightarrowtail & \mathrm{Ext}(F_o \otimes B,C) & \twoheadrightarrow & \mathrm{Ext}(\mathrm{Tor}(A,B),C) \\ \downarrow{\scriptstyle \tau} & & \downarrow{\scriptstyle \tau_o} & & \downarrow{\scriptstyle \mu'} \\ \mathrm{Hom}(F,\mathrm{Ext}(B,C)) & \rightarrowtail & \mathrm{Hom}(F_o,\mathrm{Ext}(B,C)) & \twoheadrightarrow & \mathrm{Ext}(A,\mathrm{Ext}(B,C)) \end{array}$$

The top row is exact because $\mathrm{Ext}(\_,C)$ is right exact. Since F and F_o are projective, we have that $\mathrm{Ext}(F,\mathrm{Ext}(B,C)) = 0$ and, hence, the bottom row is exact. Since F and F_o are projective, it follows from a preceding lemma that τ and τ_o are isomorphisms. Thus, μ' is an isomorphism from the 5-lemma [14].

Since $\mathrm{Hom}(A,\mathrm{Hom}(B,C)) \approx \mathrm{Hom}(B,\mathrm{Hom}(A,C))$ under the natural isomorphism which sends f: $A \to \mathrm{Hom}(B,C)$ into the map g: $B \to \mathrm{Hom}(A,C)$ such that $(g(b))(a) = (f(a))(b)$, one may use the preceding method to prove the following theorem.

<u>Theorem</u>. (Nunke) There is a natural homomorphism

$$\sigma\colon \mathrm{Ext}(A,\mathrm{Hom}(B,C)) \to \mathrm{Hom}(B,\mathrm{Ext}(A,C))$$

which is an isomorphism when B is projective, and an isomorphism

$$\nu\colon \mathrm{Ext}(A,\mathrm{Ext}(B,C)) \to \mathrm{Ext}(B,\mathrm{Ext}(A,C))$$

such that, for any exact sequence $B' \rightarrowtail B \twoheadrightarrow B''$, commutativity holds in the diagram

$$\begin{array}{ccc} \mathrm{Ext}(A,\mathrm{Hom}(B',C)) & \xrightarrow{\mathrm{Ext}(A,\delta)} & \mathrm{Ext}(A,\mathrm{Ext}(B'',C)) \\ \downarrow{\scriptstyle \sigma'} & & \downarrow{\scriptstyle \nu} \\ \mathrm{Hom}(B',\mathrm{Ext}(A,C)) & \xrightarrow{\Delta} & \mathrm{Ext}(B'',\mathrm{Ext}(A,C)) \end{array}$$

where $\delta\colon \mathrm{Hom}(B',C) \to \mathrm{Ext}(B'',C)$ and Δ are the appropriate connecting homomorphisms.

The proof is similar to that of the preceding theorem using first a projective resolution of A and then an injective resolution of C.

If $Z \rightarrowtail M \twoheadrightarrow H$ is exact, we note in the first of the above theorems that $\tau'\colon \mathrm{Ext}(Z \otimes B,C) \to \mathrm{Hom}(Z,\mathrm{Ext}(B,C))$ is a natural isomorphism which induces the commutative diagram in Theorem 76

$$\begin{array}{ccc} \mathrm{Ext}(B,C) & \xrightarrow{\mathrm{Ext}(\delta,C)} & \mathrm{Ext}(\mathrm{Tor}(H,B),C) \\ & {\scriptstyle \delta}\searrow & \downarrow{\scriptstyle \mu'} \\ & & \mathrm{Ext}(H,\mathrm{Ext}(B,C)). \end{array}$$

Furthermore, the natural isomorphism σ':

$$\sigma'\colon \mathrm{Ext}(A,\mathrm{Hom}(Z,C)) \rightarrowtail\!\!\twoheadrightarrow \mathrm{Hom}(Z,\mathrm{Ext}(A,C))$$

and the second of the above theorems yields a commutative diagram

$$\begin{array}{ccc} \mathrm{Ext}(A,C) & \xrightarrow{\mathrm{Ext}(A,\delta)} & \mathrm{Ext}(A,\mathrm{Ext}(H,C)) \\ & {\scriptstyle \Delta}\searrow & \downarrow{\scriptstyle \nu} \\ & & \mathrm{Ext}(H,\mathrm{Ext}(A,C)) \end{array}$$

establishing the commutative diagram in Theorem 77.

BIBLIOGRAPHY

1. J. W. Armstrong, "On the indecomposability of torsion-free abelian groups", Proc. Amer. Math. Soc., 16 (1965), 323-325.

2. ________, "On p-pure subgroups of the p-adic integers", in Topics in Abelian Groups, Scott, Foresman and Co., Chicago, Ill., 1963; pp. 315-321.

3. S. Balcerzyk, "On factor groups of some subgroups of a complete direct sum of infinite cyclic groups", Bull. Acad. Polon. Sci. Math. Astr. Phys., 7 (1959), 141-142.

4. R. Baer, "The decomposition of abelian groups into direct summands", Quart. J. Math., 6 (1935), 217-221.

5. ________, "Types of elements and the characteristic subgroups of abelian groups", Proc. Lon. Math. Soc., 39 (1935), 481-514.

6. ________, "The subgroup of elements of finite order of an abelian group", Ann. of Math., 37 (1936), 766-781.

7. ________, "Primary abelian groups and their automorphisms", Amer. J. of Math., 59 (1937), 99-117.

8. ________, "Abelian groups without elements of finite order", Duke Math. J., 3 (1937), 68-122.

9. ________, "Abelian groups that are direct summands of every containing abelian group", Bull. Amer. Math. Soc., 46 (1940), 800-806.

10. ________, "A unified theory of projective spaces and finite abelian groups", Trans. Amer. Math. Soc., 52 (1942), 283-343.

11. ________, "Automorphism rings of primary abelian operator groups", Ann. of Math., 44 (1943), 192-227.

12. ________, "Die Torsionsuntergruppe einer abelschen Gruppe, Math. Ann., 135 (1958), 219-234.

13. R. A. Beaumont and R. S. Pierce, "Torsion-free groups of rank two, Mem. Amer. Math. Soc., No. 38 (1961), 41 pp.

14. H. Cartan and S. Eilenberg, Homological Algebra, Princeton, 1956.

15. S. Chase, "Direct products of modules", Trans. Amer. Math. Soc., 97 (1960), 457-473.

16. ________, "Function topologies on abelian groups", Ill. J. Math., 7 (1963), 593-608.

17. __________, "On group extensions and a problem of J. H. C. Whitehead", in _Topics in Abelian Groups_, Scott, Foresman and Co., Chicago, Ill., 1963, pp. 173-193.

18. __________, "Locally free modules and a problem of Whitehead", _Ill. J. Math._, 6 (1962), 682-699.

19. A. L. S. Corner, "A note on rank and direct decompositions of torsion-free abelian groups", _Proc. Camb. Phil. Soc._, 57 (1967), 230-233.

20. __________, "Every countable reduced torsion-free ring is an endomorphism ring", _Proc. Lon. Math. Soc._, 13 (1963), 687-710.

21. __________, "On a conjecture of Pierce concerning direct decompositions of abelian groups", _Proc. of Coll. on Abelian Groups_, Budapest, 1964, pp. 43-48.

22. __________, "Endomorphism algebras of large modules with distinguished submodules", to appear.

23. A. L. S. Corner and P. Crawley, "An abelian p-group without the isomorphic refinement property", to appear.

24. P. Crawley, "Solution of Kaplansky's test problems for primary abelian groups", _J. Alg._, 2 (1965), 413-431.

25. __________, "The cancellation of torsion abelian groups in direct sums", _J. Alg._, 2 (1965), 432-442.

26. __________, "An isomorphic refinement theorem for certain abelian p-groups", _J. Alg._, 6 (1967), 376-387.

27. __________, "Abelian p-groups determined by their Ulm sequences", _Proc. J. of Math._, 22 (1967), 235-239.

28. P. Crawley and A. W. Hales, "The structure of torsion abelian groups given by presentations", _Bull. Amer. Math. Soc._, 74 (1968), 954-956.

29. DeGroot, "Indecomposable abelian groups", _Proc. Nederl. Akad. Wet._, 60 (1959), 137-145.

30. J. Erdos, "On the splitting problem of mixed abelian groups", _Publ. Math. Debrecen_, 5 (1958), 364-377.

31. S. Fomin, "Uber peridische Untergruppen der undlichen abelschen Gruppen", _Mat. Sbornik_, 2 (1937), 1007-1009.

32. L. Fuchs, _Abelian Groups_, Budapest, 1958.

33. __________, "Notes on abelian groups I", _Ann. Univ. Sci. Budapest Eötvös Sect. Math._, 2 (1959), 5-23.

34. __________, "On character groups of discrete abelian groups", _Acta Math. Acad. Sci. Hungar._, 10 (1959), 133-140.

35. __________, "The existence of indecomposable abelian groups of arbitrary power", _Acta Math. Acad. Sci. Hungar._, 10 (1959), 453-457.

36. __________, "Notes on abelian groups II, *Acta Math. Acad. Sci. Hungar.*, 11 (1960), 117-125.

37. __________, "Recent results and problems on abelian groups", in *Topics in Abelian Groups*, Scott, Foresman and Co., 1963, pp. 9-40.

38. __________, "Note on purity and algebraic compactness for modules", to appear.

39. P. Griffith, "Purely indecomposable torsion free groups", *Proc. Amer. Math. Soc.*, 18 (1967), 738-742.

40. __________, "A counterexample to a theorem of Chase", *Proc. Amer. Math. Soc.*, 19 (1968), 923-924.

41. __________, "Decompositions of pure subgroups of torsion free groups", *Ill. J. Math.*, 12 (1968), 433-438.

42. __________, "Separability of torsion free groups and a problem of J. H. C. Whitehead", *Ill. J. Math.*, 12 (1968), 654-659.

43. __________, "A solution to the splitting mixed group problem of Baer", *Trans. Amer. Math. Soc.*, 139 (1969), 261-269.

44. __________, "On direct sums of p-mixed groups", *Archiv Der Mathematik*, 19 (1968), 359-360.

45. __________, "A note on a theorem of Hill", *Pac. J. Math.*, 29 (1969), 279-284.

46. __________, "Extensions of free groups by torsion groups", *Proc. Amer.* Math. Soc., to appear.

47. D. K. Harrison, "Infinite abelian groups and homological methods", *Ann. of Math.*, 69 (1959), 366-391.

48. P. Hill, "Pure subgroups having prescribed socles", *Bull. Amer. Math.* Soc., 71 (1965), 608-609.

49. __________, "Sums of countable primary groups", *Proc. Amer. Math. Soc.*, 17 (1966), 1469-1470.

50. __________, "Quasi-isomorphism of primary groups", *Mich. Math. J.*, 13 (1966), 481-484.

51. __________, "The isomorphic refinement theorem for direct sums of closed groups", *Proc. Amer. Math. Soc.*, 18 (1967), 913-919.

52. __________, "On primary groups with uncountably many elements of infinite height", *Archiv Math.*, 19 (1968), 279-283.

53. __________, "Isotype subgroups of direct sums of countable groups", *Ill. J. Math.*, to appear.

54. __________, "On the classification of abelian groups", to appear.

55. __________, "Extending automorphisms on primary groups", *Bull. Amer. Math. Soc.*, 74 (1968), 1123-1124.

56. _________, "The purification of subgroups of abelian groups", submitted for publication.

57. _________, "On the decomposition of groups", submitted for publication.

58. _________, "A countability condition for primary groups presented by relations of length two", Bull. Amer. Math. Soc., 75 (1969), 780-782.

59. _________ and C. Megibben, "Minimal pure subgroups in primary groups", Bull. Soc. Math. France, 92 (1964), 251-257.

60. _________ and C. Megibben, "Quasi-closed primary groups", Acta Math. Acad. Sci. Hungar., 16 (1965), 271-274.

61. _________ and C. Megibben, "Extending automorphisms and lifting decompositions in abelian groups", Math. Annalen, 175 (1968), 159-168.

62. _________ and C. Megibben, "On primary groups with countable basic subgroups", Trans. Amer. Math. Soc., 124 (1966), 49-59.

63. _________ and C. Megibben, "On direct sums of countable groups and generalizations", in Etudes sur les Groupes Abeliens, Springer-Verlag, 1967, pp. 183-206.

64. A. Hulanicki, "Algebraic structure of compact abelian groups", Bull. Acad. Pol. Sci., 6 (1958), 71-73.

65. J. Irwin, "High subgroups of abelian torsion groups", Pac. J. Math., 11 (1961), 1375-1384.

66. J. Irwin and E. A. Walker, "On isotype subgroups of abelian groups", Bull. Soc. Math.

67, B. Jónsson, "On direct decomposition and torsion-free abelian groups", Math. Scand., 5 (1957), 230-235.

68. I. Kaplansky, "Elementary divisors and modules", Trans. Amer. Math. Soc., 66 (1949), 464-491.

69. _________, Infinite Abelian Groups, Michigan Press, Ann Arbor, 1969.

70. _________, "Modules over Dedekind rings and valuation rings", Trans. Amer. Math. Soc., 72 (1952), 327-340.

71. _________, "Projective modules", Ann. of Math., 68 (1958), 372-377.

72. _________, "Decomposability of modules", Proc. Amer. Math. Soc., 13 (1962), 532-535.

73. _________, "The splitting of modules over integral domains", Archiv. Math., 13 (1962), 341-343.

74. I. Kaplansky and G. W. Mackey, "A generalization of Ulm's Theorem", Summa Brasil Math., 2 (1951), 195-202.

75. G. Kolettis, "Direct sums of countable groups", Duke Math. J., 27 (1960), 111-125.

76. __________, "Semi-complete primary abelian groups", Proc. Amer. Math. Soc., 11 (1960), 200-205.

77. __________, "Homogeneously decomposable modules", in Etudes sur les Groupes Abeliens, Springer-Verlag, New York, 1962, pp. 223-238.

78. G. Kothe, "Verallgemeinerte abelsche Gruppen mit hyperkomplexem Operatorenring", Math. Zeit., 39 (1935), 31-44.

79. L. Kulikov, "Zur Theorie der abelschen Gruppen von beliebiger Mächtigheit", Mat. Sbornik, 9 (1941), 165-181.

80. __________, "On the theory of abelian groups of arbitrary power", Mat. Sbornik, 16 (1945), 129-162.

81. __________, "Generalized primary groups I", Trudy Moskov Mat. Obsc., 1 (1952), 247-326. II - Same Trudy 2 (1953), 85-167.

82. __________, "On direct decompositions of groups", Ukrain Mat. Z., 4 (1952), 230-275, 347-372.

83. A. Kurosh, "Primitive torsionfreie abelsche Gruppen von endlichen Range", Ann. of Math., 38 (1937), 175-203.

84. __________, Theory of Groups, second edition, Moscow, 1953, English translation, Chelsea, New York, 1955-1956.

85. H. Leptin, "Abelsche p-Gruppen und ihre Automorphismengruppen", Math. Z., 73 (1960), 235-253.

86. __________, "Zur Theorie die überabzählbaren abelschen p-Gruppen", Abhandl. Math. Sem. Univ. Hamb., 24 (1960), 79-90.

87. F. W. Levi, "Abelsche Gruppen mit abzahlbaren Elementen", Leipzig, 1917.

88. S. MacLane, Homology, Berlin: Springer-Verlag OHG; New York: Academic Press, 1963.

89. A. I. Malcev, "Torsionfreie abelsche Gruppen von endlichen Range", Mat. Sbornik, 4 (1938), 46-58.

90. J. M. Maranda, "On pure subgroups of abelian groups", Arch. der Math., 11 (1960), 1-13.

91. E. Matlis, "Injective modules over Noetherian rings", Pac. J. Math., 8 (1958), 511-528.

92. __________, "Divisible modules", Proc. Amer. Math. Soc., 11 (1960), 385-391.

93. __________, "Cotorsion modules", Mem. Amer. Math. Soc., No. 49, 1964.

94. __________, "The decomposability of torsion-free modules of finite rank", Trans. Amer. Math. Soc., to appear.

95. C. Megibben, "Large subgroups and small homomorphisms", Mich. Math. J., 13 (1966), 153-160.

96. __________, "On high subgroups", Pac. J. Math., 14 (1964), 1353-1358.

97. __________, "On mixed groups of torsion-free rank one", Ill. J. Math., 11 (1967), 134-143.

98. __________, "Modules over an incomplete discrete valuation ring", Proc. Amer. Math. Soc., 19 (1968), 450-452.

99. __________, "The generalized Kulikov criterion", submitted for publication.

100. __________, "A generalization of the classical theory of primary groups", submitted for publication.

101. __________, "A nontransitive, fully transitive primary group", J. Alg., to appear.

102. R. Nunke, "Modules of extensions over Dedekind rings", Ill. J. Math., 3 (1959), 222-241.

103. __________, "Purity and subfunctors of the identity", in Topics in Abelian Groups, Scott, Foresman and Co., Chicago, Ill., 1963, pp. 121-171.

104. __________, "On the structure of Tor", in Proc. of Coll. on Abelian Groups, Budapest, 1964, pp. 115-124.

105. __________, "Homology and direct sums of countable abelian groups", Math Zeit., 101 (1967), 182-212.

106. __________, "On the structure of Tor II", Pac. J. Math., 22 (1967), 453-464.

107. __________, "Slender groups", Acta Sci. Math. Szeged, 23 (1962), 67-73.

108. Z. Papp, "On algebraically closed modules", Publ. Math. Debrecen, 6 (1959), 311-327.

109. L. D. Parker and E. A. Walker, "An extension of the Ulm-Kolettis theorems", in Etudes sur les Groupes Abeliens, Springer-Verlag, New York, 1968, pp. 309-325.

110. R. S. Pierce, "Homomorphisms of primary abelian groups", in Topics in Abelian Groups, Scott, Foresman and Co., Chicago, Ill., 1963, pp. 215-230.

111. __________ and R. Beaumont, "Quasi-isomorphism of p-groups", in Proc. of Coll. on Abelian Groups, Budapest, 1964, pp. 13-27.

112. H. Prufer, "Untersuchungen uber die Zerlegbarkeit der abzahlbaren primaren abelschen Gruppen", Math. Zeit., 17 (1923), 35-61.

113. __________, "Theorie der abelschen Gruppen I, Grundeigenschaften, Math. Zeit., 20 (1924), 165-87; II, Ideale Gruppen, Math. Zeit., 22 (1925), 222-249.

114. F. Richman and E. A. Walker, "Extending Ulm's theorem without group theory", to appear.

115. __________ and J. Irwin, "Direct sums of countable groups and related concepts", J. Alg., 2 (1965), 443-450.

116. J. Rotman, "Mixed modules over valuation rings", Pac. J. Math., 10 (1960), 607-623.

117. __________, "On a problem of Baer and a problem of Whitehead in abelian groups", Acta Math. Acad. Sci. Hungar., 12 (1961), 245-254.

118. __________, "A characterization of fields among integral domains", Proc. Amer. Math. Soc., 11 (1960), 356-360.

119. __________, "A note on completions of modules", Proc. Amer. Math. Soc., 11 (1960), 356-360.

120. __________, The Theory of Groups: An Introduction, Allyn and Bacon, Boston, 1965.

121. __________ and T. Yen, "Modules over a complete discrete valuation ring", Trans. Amer. Math. Soc., 98 (1961), 242-254.

122. E. Sasida, "Negative solution of I. Kaplansky's first test problem for abelian groups and a problem of K. Borsuk concerning cohomology groups", Bull. Acad. Polon., 9 (1961), 331-334.

123. __________, "Proof that every countable and reduced torsion free group is slender".

124. __________, "On abelian groups every countable subgroup of which is an endomorphic image", Bull. Acad. Polon. Sci. Cl. III, 2 (1954), 359-362.

125. __________, "An application of Kulikov's basic subgroups in the theory of abelian mixed groups", Bull. Acad. Polon. Sci. Cl. III, 4 (1956), 411-413.

126. __________, "Construction of a direct indecomposable abelian group of power higher than that of the continuum", Bull. Acad. Polon. Sci. Cl. III, 5 (1957), 701-703.

127. __________, "Uber die characteristicher Untergruppen einer endlichen abelschen Gruppe", Math. Zeit., 31 (1960), 611-624.

128. E. Specker, "Additive Gruppen von Folgen ganzer Zahlen", Portugaliae Math., 9 (1950), 131-140.

129. K. Stein, "Anslystische Funktionen Mehrerer Komplexer Veranderlichen zu Vorgegebenen Periodizitatsmodulum und das Zweite Cousinsche Problem, Math. Ann., 123 (1951), 201-222.

130. H. Ulm, "Zur Theorie der abzahlbar-unedlichen abelschen Gruppen", Math. Ann., 107 (1933), 774-803.

131. __________, "Elementarteilertheorie unendlichen Matrixen", Math. Ann., 114 (1937), 493-505.

132. C. Walker, "Relative homological algebra and abelian groups", Ill. J. Math., 10 (1966), 186-209.

133. ________, "Properties of Ext and quasi-splitting of abelian groups", _Acta Math. Acad. Sci. Hungar._, 15 (1964), 157-160.

134. E. A. Walker, "Cancellation in direct sums of groups", _Proc. Amer. Math. Soc._, 7 (1956), 898-902.

135. ________, F. Richman and C. Walker, "Projective classes of abelian groups", in _Etudes sur les Groupes Abeliens_, Springer-Verlag, New York, 1968, pp. 335-343.

136. ________, F. Richman and C. Walker, "On p^{α}-pure sequences of abelian groups", in _Topics in Abelian Groups_, Scott, Foresman and Co., Chicago, Ill., 1963, pp. 69-119.

137. R. B. Warfield, "Purity and algebraic compactness for modules", _Pac. J. Math._, 28 (1969), 699-719.

138. L. Zippen, "Countable torsion groups", _Ann. of Math._, 36 (1935), 86-99.

INDEX

37-100